Trip Generation Handbook

Second Edition

An ITE Recommended Practice

June 2004

Institute of Transportation Engineers

Trip Generation Handbook, Second Edition

A Recommended Practice
of the Institute of Transportation Engineers

Principal Editor:
Kevin G. Hooper, P.E.

Oversight and technical review
of this publication were provided by the
Trip Generation Advisory Committee:
Chair, Eugene D. Arnold Jr., P.E.

This version of the *Trip Generation Handbook*, Second Edition, RP–028B, incorporates changes necessary for consistency with the data contained in the Seventh Edition of *Trip Generation*, which was released in November 2003. The recommendations in this publication have not changed from the 2001 edition of the handbook. Additional data have been added to Chapter 5, Pass-By, Primary and Diverted Linked Trips. All other changes were strictly editorial updates to the material contained in the 2001 handbook.

The *Trip Generation Handbook*, RP–028A, was approved in November 2000 as a recommended practice of the Institute of Transportation Engineers (ITE). It supersedes the proposed recommended practice, RP–028, dated October 1998. The comment period on the proposed recommended practice closed on January 1, 2000. Comments on the October 1998 document have been incorporated into this document.

Certain individual volunteer members of ITE's recommended practice development bodies are employed by Federal agencies, other governmental offices, private enterprise, or other organizations. Their participation in ITE recommended practice development activities does not constitute endorsement by these government agencies or other organization endorsement of any of ITE recommended practice development bodies or any ITE recommended practices that are developed by such bodies.

The Institute of Transportation Engineers is an international educational and scientific association of transportation and traffic engineers and other professionals who are responsible for meeting mobility and safety needs. ITE facilitates the application of technology and scientific principles to research, planning, functional design, implementation, operation, policy development, and management for any mode of transportation by promoting professional development of members, supporting and encouraging education, stimulating research, developing public awareness, and exchanging professional information; and by maintaining a central point of reference and action.

Founded in 1930, ITE serves as a gateway to knowledge and advancement through meetings, seminars and publications; and through a network of some 16,000 members working in some 90 countries. ITE also has more than 70 local and regional chapters and more than 120 student chapters that provide additional opportunities for information exchange, participation and networking.

Institute of Transportation Engineers
1099 14th Street, NW, Suite 300 West, Washington, DC 20005-3438 USA
Telephone: +1 202-289-0222, Fax: +1 202-289-7722
ITE on the Web: www.ite.org

Table of Contents

Preface

The Institute of Transportation Engineers (ITE) published the Seventh Edition of *Trip Generation* in three volumes in 2003. The first volume, the *User's Guide*, describes the database used in the next two volumes, as well as the data plots and their reported statistics. The second and third volumes contain land use descriptions and data plots. *Trip Generation* is an ITE informational report.

During the development of *Trip Generation*, ITE decided to separate the processing and dissemination of the informational trip generation data from the development of recommendations on how to use and apply the data. This separate document, the *Trip Generation Handbook*, 2nd Edition, is a recommended practice of ITE.

The *Trip Generation Handbook* has two primary purposes: to provide instruction and guidance in the proper use of data presented in *Trip Generation*, and to provide information on supplemental issues of importance in estimating trip generation for development sites.

Because the instruction and guidance in the main body of this handbook represents a recommended practice for estimating trip generation, its function is distinct from that of *Trip Generation*. Note, however, the data contained in the appendices of this handbook are for informational purposes only.

The analysis methods presented in the handbook have been developed to be simple and understood by the novice transportation planner/engineer, yet sufficiently accurate for the experienced transportation professional.

Acknowledgments

This handbook is the result of many months of concerted effort by dedicated volunteers, both ITE members and non-members, and the ITE headquarters staff.

ITE Headquarters retained an editor, Kevin G. Hooper, to coordinate the development of the recommended practice. The editor was responsible (in general) for assembling the existing materials on the proposed contents of the handbook, editing and revising these materials, assembling the draft handbook and coordinating the standard procedures for developing the handbook as a recommended practice. ITE is most appreciative of the tireless efforts and excellent work performed by Kevin Hooper.

ITE is also particularly appreciative of efforts by Eugene D. Arnold Jr. and Brian S. Bochner. Their dedicated service, expertise and insight contributed immensely to the completion of this recommended practice.

Oversight and technical review during the development of this handbook was provided by the Trip Generation Advisory Committee:

Eugene D. Arnold Jr., P.E. (F), Chair
Brian S. Bochner, P.E. (F)
Lisa M. Fontana Tierney (M)

Kenneth G. Mackiewicz, P.E. (M)
David Muntean, P.E. (M)
Allyn D. Rifkin, P.E. (M)

Endorsement for publication of this handbook as a Recommended Practice of ITE was provided by the Trip Generation Standards Review Panel consisting of the following individuals:

John J. DeShazo, P.E. (F)
Kathryn Z. Heffernan, P.E. (M)
Robert P. Jurasin, P.E. (F)

Thomas E. Mitchell, P.E. (M)
Robert C. Wunderlich, P.E. (M)

Approval to publish this handbook as a proposed recommended practice of ITE was provided by the 1998 ITE International Board of Direction. Approval to publish this handbook as a recommended practice of ITE was provided by the 2000 ITE International Board of Direction.

Reviewers of the draft procedures and guidelines during preparation of the draft proposed recommended practice included the following individuals:

Ahmad D. Al-Akhras, P.E. (M)
David E. Clark, P.E. (A)
Laura B. Firtel, AICP (A)
Jon D. Fricker, P.E. (M)
Patrick A. Gibson, P.E. (F)
James W. Gough, P.E. (M)
Richard W. Lyles, P.E. (M)

Christopher B. Middlebro, P.E. (M)
Robert P. Murphy, P.E. (M)
David S. Plummer, P.E. (M)
Steven A. Tindale, P.E. (M)
Robert P. Wallace, AICP (A)
Robert M. Winick (M)

Initial drafts of the handbook chapters were prepared by the following individuals:

Ahmad D. Al-Akhras, P.E. (M)
Brian S. Bochner, P.E. (F)
Robert T. Dunphy (A)
Kevin G. Hooper, P.E. (M)

Richard W. Lyles, P.E. (F)
Kenneth G. Mackiewicz, P.E. (M)
Robert E. Stammer, P.E. (F)

CHAPTER 1
Introduction

1.1 Use of *Trip Generation* Data

The thousands of data points presented in *Trip Generation*, 7th Edition are used for a variety of purposes. For example, developers determine **site access requirements** for proposed developments with the aid of *Trip Generation*. Traffic engineers use trip generation data to estimate future traffic volumes upon which **off-site transportation improvements** are based. Communities use the data to evaluate the implications of requests for **zoning changes** or of potential **land use changes.**

The complexity of potential trip generation issues has grown for the novice transportation planner, local planning officials and the veteran of *Trip Generation* starting back in its earliest editions. The many issues facing the users of trip generation data include the following:

◆ Need to select an appropriate trip generation estimation method when *Trip Generation* data (1) includes both a weighted average rate and a regression equation or (2) is based on only a limited sample of data points;

◆ Need to collect local trip generation data to either validate the use of *Trip Generation* data for local use or establish a new local trip generation rate;

> **Due to the widespread use of *Trip Generation*, it is important that ITE promote proper application and provide guidelines regarding appropriate interpretation of the data.**

◆ Increased recognition of the pass-by trip-making phenomenon as a significant factor for some land uses;

◆ Continual evolution of the mixes of land use, site densities and on-site amenities within land development (e.g., multi-use developments);

◆ Occasional need to estimate trip generation for a site before the land use mix or density of a development are known;

◆ Introduction of new issues (e.g., how many truck trips are generated); and

◆ Claim that specific transportation demand management programs and transit services will reduce site trip generation by a certain amount.

One issue that has not changed is the need for ethics and objectivity in the use of *Trip Generation* data. Although study preparers and reviewers may have different objectives and perspectives, all parties involved in the development of trip generation estimates should adhere to established engineering ethics similar to the ITE Canon of Engineering Ethics and conduct all analyses and reviews objectively, accurately and professionally.

1.2 Purpose of the *Trip Generation* Handbook

The *Trip Generation Handbook*, 2nd Edition, has two primary purposes. The first is to provide instruction and guidance in the proper use of data presented in *Trip Generation*. The second is to provide additional information and data on supplemental issues of importance in estimating trip generation for development sites.

The instruction and guidance in the main body of this handbook represent a recommended practice for estimating trip generation for a development site. As such, its function is distinct from that of *Trip Generation*, which is an ITE informational report.

The analysis methods presented in the handbook have been developed to be simple and understandable to the novice transportation planner/ engineer, while being sufficiently accurate for the experienced transportation professional.

1.3 Contents of the *Trip Generation Handbook*

Chapter 2 provides guidance in the selection of the appropriate **independent variable** and **time period** for estimating trips generated by a development site.

Chapter 3 presents a recommended methodology for **estimating trip generation** for a land use for which data are provided in *Trip Generation.*

Chapter 4 offers guidelines for the proper execution of a **local trip generation study,** for the validation of *Trip Generation* data for local use, or for the establishment of new local trip generation rates.

Chapter 5 explains the concepts of **pass-by, primary and diverted linked trips** and their importance

in estimating trips at a site's driveway(s); presents a data base on pass-by trips compiled by ITE; proposes a recommended methodology for interpreting and applying the pass-by data; and provides instructions on how to conduct a pass-by survey (with sample size requirements, data requirements and a sample survey instrument).

Chapter 6 presents an approach for estimating trips generated by a **proposed site with an unknown mix of land uses or size of development.**

Chapter 7 includes a definition of **multi-use developments** and presents a method for estimating internal capture rates (and the resultant reduction in external trips) within a multi-use development.

The handbook also contains a series of appendices that provide information on additional topics related to trip generation. The information provided in the appendices is strictly informational and is not intended to recommend practices or procedures.

Appendix A presents information on **truck trip generation** data.

Appendix B presents information on the potential effects of **transportation demand management** programs and transit availability on trip generation rates.

Appendix C summarizes key **literature on multi-use development** that was used in the preparation of the Chapter 7 guidelines.

Appendix D presents a **glossary of terms** used throughout this handbook.

CHAPTER 2

Selection of Independent Variable and Time Period for Analysis

2.1 Definitions of Independent Variables

For the purposes of estimating trip generation, an independent variable is defined as a physical, measurable and predictable unit describing the study site or trip generator (e.g., gross floor area, employees, seats, dwelling units). *Trip Generation* presents, for each land use, the independent variable or variables that appear to be a "cause" for the variation in the number of trip ends generated by a land use.

It is critical that the analyst understand the definition of each potential independent variable for a particular land use. The analyst should carefully read the *Trip Generation* definitions for all independent variables being considered (note: some definitions are presented in Chapter 3, Volume 1, *User's Guide*, of *Trip Generation*, Seventh Edition; the glossary of this handbook also presents a comprehensive listing of independent variable definitions).

If the analyst has reason to believe that the independent variable (and how it was measured for sites reported in *Trip Generation*) does not match the characteristics of a site under analysis, a local trip generation study should be conducted (see Chapter 4) or appropriate refinements made to achieve consistency.

2.2 Selection of Independent Variable, If a Choice Is Available

For many land uses presented in *Trip Generation*, vehicle trip generation rates and equations have been provided for more than one independent variable. The choice of variable can be one of the most important decisions in calculating trip generation. Sometimes there is no choice because the information available for the site under study relates only to a single independent variable.

Preferred Independent Variable

◆ Appears to be a "cause" for the variation in trip ends generated by a land use;

◆ Obtained through a primary measurement and not derived from secondary data;

◆ Produces a rate/equation with the "best fit" of data;

◆ Can be reliably forecast for applications; and

◆ Related to the land use type and not solely to the characteristics of the site tenants.

When the analyst has a choice of variables, it is best to use the one that (1) is most **directly causal for the variation in trip ends** generated by a land use and (2) is **accurately projectable** for proposed development sites. Correlation coefficients between the vehicle-trips measured (e.g., average week-day trips) and independent variables are provided with the data plots. The standard deviation and the coefficient of determination (R^2) values indicate which independent variable **best fits the data.** Standard deviations less than or equal to 110 percent of the weighted average rate, and R^2 values of 0.75 or greater, are both indicative of good fits with the data. (Note: a discussion of statistical terms, such as R^2, is found in *Trip Generation, User's Guide*, Seventh Edition, Volume 1 and in Appendix D of this handbook.)

It is also important to check the sample size for each independent variable. In the case of two variables with similar measures of "best fit," the analyst should usually favor the most accurately projected variable. If there appears to be little difference, then the variable with the larger sample size should be favored.

The preferred independent variable should be stable for a particular land use type and not a direct function of actual site tenants. In other words, the values and measurements attributable to an independent variable should not change dramatically with changes in building tenants. Physical site characteristics (e.g., square feet of floor area, number of dwelling units) are preferable.

Finally, the best independent variable is **obtained through a primary measurement,** not derived from secondary data. For example, many estimates of the number of employees working in an office building are derived as a function of the size (in square footage) of the office building and an assumed employment density. This approach is not likely to be accurate. In such a case, the preference should be to use the office building square footage as the primary independent variable.

2.3 Selection of Independent Variable, If the Measure Must Be Derived

In the planning stage, some independent variables may need to be estimated on the basis of other variables. For example, the amount of employment is generally estimated on the basis of gross floor area (GFA). Therefore, GFA would be the strongest variable.

Preference should be given to independent variables that are directly available. If an estimate is needed, use a different independent variable known to be valid and accurate and apply a realistic and credible factor to generate the desired independent variable. Such estimates should be based on verifiable or valid relationships applicable to the site being considered. It is always best to review such projections or conversions with the reviewer of the estimate to gain early consensus. Otherwise, all subsequent work may have to be redone due to the invalid conversion.

2.4 Time Period for Analysis

Trip Generation is intended for use in estimating the number of trip ends that may be generated by a specific land use. Selection of the time period for a trip generation study is dictated by the purpose of a traffic impact assessment for which the estimate is being made.

To determine the appropriate traffic impacts and resulting design requirements, the analyst should examine the weighted average rates or regression equations for the different days and time periods to determine when the site in question peaks in traffic generation. It is also critical to define the relationship between that peak generation and the peaking characteristics of the adjacent street system. **The time period that should be analyzed is the time period in which the combination of site-generated traffic and adjacent street traffic is at its maximum.**

For nearly all cases, the generation rates or equations for the morning and evening peak hours of the adjacent street system (highest one-hour period between 7:00 a.m. and 9:00 a.m. and between 4:00 p.m. and 6:00 p.m.) would be used to test the impact on the normal peak hour traffic. Some land uses, however, do not peak at the same time as the adjacent streets (e.g., theaters, factory shift that ends at 3 p.m.). Therefore, the analyst should test combinations of generator volumes and street

volumes at different times to determine a site's maximum, and most critical, impact.

Trip generation data plots, rates and equations are presented in *Trip Generation* for a range of days of the week (i.e., average weekday, Saturday, Sunday) and for different time periods during those days (i.e., the daily morning and evening peak hours of the generator and of adjacent street traffic).

The ITE recommended practice *Traffic Access and Impact Studies for Site Development* provides further guidance on defining the appropriate time period.

2.5 Time of Day, Day of Week and Seasonal Variations

Tables 2.1 through 2.4 (reproduced from *Trip Generation*, Seventh Edition) provide miscellaneous information on daily, monthly and hourly variations in shopping center (Land Use Code 820) traffic. It should be noted, however, that the number of studies providing these data are limited, and caution in using these tables is therefore recommended. **Trip generation for the peak hour should not be determined from the hourly variation information provided in Tables 2.1 and 2.2. Peak hour trip generation volumes should be derived directly from the data plots contained in ITE's *Trip Generation*.**

Table 2.1
Hourly Variation in Shopping Center Traffic
Less Than 100,000 Square Feet Gross Leasable Area

TIME	AVERAGE WEEKDAY		AVERAGE SATURDAY	
	PERCENT OF 24-HOUR ENTERING TRAFFIC	PERCENT OF 24-HOUR EXITING TRAFFIC	PERCENT OF 24-HOUR ENTERING TRAFFIC	PERCENT OF 24-HOUR EXITING TRAFFIC
10:00–11:00 a.m.	7.6	6.5	6.8	5.8
11:00 a.m.–12:00 p.m.	7.6	8.4	8.8	8.9
12:00–1:00 p.m.	7.6	8.2	9.4	8.8
1:00–2:00 p.m.	6.9	7.5	10.0	10.1
2:00–3:00 p.m.	9.0	7.8	9.7	8.4
3:00–4:00 p.m.	9.6	9.5	10.3	9.6
4:00–5:00 p.m.	9.7	10.4	10.7	10.7
5:00–6:00 p.m.	10.3	11.0	9.4	8.7
6:00–7:00 p.m.	7.4	8.3	7.3	8.3
7:00–8:00 p.m.	5.4	5.3	5.0	5.7
8:00–9:00 p.m.	4.2	4.3	3.2	3.9
9:00–10:00 p.m.	1.9	1.8	2.0	3.3

Source: *Trip Generation*, Seventh Edition

Table 2.2
Hourly Variation in Shopping Center Traffic
More Than 300,000 Square Feet Gross Leasable Area

TIME	AVERAGE WEEKDAY		AVERAGE SATURDAY		AVERAGE SUNDAY	
	PERCENT OF 24-HOUR ENTERING TRAFFIC	PERCENT OF 24-HOUR EXITING TRAFFIC	PERCENT OF 24-HOUR ENTERING TRAFFIC	PERCENT OF 24-HOUR EXITING TRAFFIC	PERCENT OF 24-HOUR ENTERING TRAFFIC	PERCENT OF 24-HOUR EXITING TRAFFIC
10:00–11:00 a.m.	7.5	3.7	8.3	4.3	3.5	1.7
11:00 a.m.–12:00 p.m.	8.6	5.9	10.9	6.9	9.4	3.5
12:00–1:00 p.m.	9.5	7.9	11.9	8.9	15.3	6.3
1:00–2:00 p.m.	8.7	8.2	12.5	10.4	17.3	11.0
2:00–3:00 p.m.	7.9	8.8	12.4	12.0	16.4	14.4
3:00–4:00 p.m.	7.7	8.9	11.2	12.9	13.8	16.2
4:00–5:00 p.m.	8.2	9.1	9.2	13.4	9.8	16.8
5:00–6:00 p.m.	8.3	9.5	5.2	12.7	5.5	15.7
6:00–7:00 p.m.	7.8	7.7	2.9	8.0	2.2	6.1
7:00–8:00 p.m.	8.4	7.0	1.9	2.1	1.3	1.9
8:00–9:00 p.m.	4.7	7.7	1.4	1.2	0.8	1.1
9:00–10:00 p.m.	1.8	9.1	2.9	0.8	0.6	0.9

Source: *Trip Generation*, Seventh Edition

Table 2.3
Daily Variation in Shopping Center Traffic
Percentage of Average Weekday Volume (Monday through Friday)

DAY	LESS THAN 100,000 SQ. FT. GLA	100,000 TO 300,000 SQ. FT. GLA	MORE THAN 300,000 SQ. FT. GLA	DISCOUNT CENTER
Sunday	45.2	65.4	77.4	82.1
Monday	97.3	96.8	96.8	95.1
Tuesday	92.9	103.1	97.1	91.4
Wednesday	92.7	99.1	93.6	94.8
Thursday	98.2	85.3	97.1	99.5
Friday	118.9	108.7	115.4	119.2
Saturday	128.5	113.4	128.0	151.0
Sample Size	**6**	**8**	**17**	**2**

Source: *Trip Generation,* Seventh Edition

Table 2.4
Monthly Variation in Shopping Center Traffic
Percentage of Average Month

MONTH	PERCENTAGE	MONTH	PERCENTAGE
January	85.3	July	100.8
February	78.1	August	102.1
March	92.0	September	94.8
April	93.2	October	98.9
May	105.4	November	101.5
June	106.0	December	141.8

Sample Size: 2
Average Gross Leasable Area: 938,000 sq. ft.
Source: *Trip Generation,* Seventh Edition

CHAPTER 3

Guidelines for Estimating Trip Generation

3.1 Background

Trip Generation provides an abundance of data on the relationships between vehicle trip generation and site characteristics. The challenge to the analyst is to make a reasonable and responsible estimate of trips generated for a particular development under consideration.

The *User's Guide* (Volume 1 of *Trip Generation*, Seventh Edition) explains the various types of data provided in Volumes 2 and 3 (e.g., data plots, weighted average rates, standard deviations, "best fit" curves, regression equations). This chapter provides guidance in the interpretation of these data and **recommends a step-by-step procedure for developing a trip generation estimate using *Trip Generation* data.**

It should be emphasized that **selection of the rate or the equation may be dictated by local ordinance or agency policy.** In some locales, for example, the analyst may be instructed to always use the regression equation, if it is given. **However, the approach described herein is recommended over an arbitrary policy because it is sensitive to data quality and therefore is likely to be more accurate.**

> If available, properly collected and validated local rates should be considered in addition to the national database.

3.2 Available Information on Trip Generation

The data presented in *Trip Generation* allow for several types of analyses of trip generation data for each combination of land use type, independent variable and time period:

◆ **Data Plot:** The most elementary display of the available information is a plot of trip ends versus a related independent variable for each individual study (or observation). The data plot is provided when there are at least two points.

◆ **Weighted Average Trip Generation Rate:** This rate is defined as the number of weighted trip ends per unit of the independent variable. The rate simply assumes a linear relationship between trip ends and the independent variable, having a slope equal to the rate and with the straight line passing through the origin (i.e., with a value of zero for the independent variable, the number of trips generated is zero). The average rates presented in *Trip Generation* are weighted averages (weighted by the units of the independent variable).

The **standard deviation** (given when there are three or more points) is a measure of how widely dispersed the data points are around the calculated average. The less dispersion, the lower the standard deviation.

In data plots for which all of the points correspond to a single value for the independent variable, no line is drawn. One example of this phenomenon is plotted for Land Use Code 853 (Convenience Market with Gasoline Pumps), Average Vehicle Trip Ends versus Vehicle Fueling Positions on a Weekday, shown on page 1,550 of *Trip Generation*, Seventh Edition. In this example, all of the sample data points have the same number of fueling positions (i.e., four); therefore, a line corresponding to an average rate cannot be drawn.

◆ **Regression Equation:** Regression analysis provides a tool for developing an equation that defines the line that "best fits" the data. This specific mathematical relationship between trip ends and the related independent variable is defined as the regression equation.

The coefficient of determination (R^2) is an estimate of the accuracy of the fit. It is the percent of the variance in the number of the trips explained by the variance in the independent variable. Thus, an R^2 of 0.64 indicates that 64 percent of the variance in the number of trips is accounted for by the variance in the independent variable. Therefore, the closer the R^2 value is to 1.0, the stronger the relationship between the number of trips and the independent variable.

The best-fit curve and an equation are shown in *Trip Generation*, Seventh Edition when there are four or more data points and when the R^2 is greater than or equal to 0.50.

Table 3.1 summarizes the types of mathematical and statistical information provided in *Trip Generation*, Seventh Edition as a function of the number of available data points. The table also indicates that, for data sets with five or fewer data points, the statement "Caution—use carefully—small sample size" is provided.

Table 3.1
Information Provided in *Trip Generation*, Seventh Edition as a Function of Data Sample Size

NUMBER OF DATA POINTS	WEIGHTED AVERAGE RATE AND RANGE OF RATES	DATA PLOT	STANDARD DEVIATION	REGRESSION EQUATION IF $R^2 \geq 0.5$	CAUTION REGARDING SMALL SAMPLE SIZE
1	Yes[1]	No	No	No	Yes
2	Yes	Yes	No	No	Yes
3	Yes	Yes	Yes	No	Yes
4	Yes	Yes	Yes	Yes	Yes
5	Yes	Yes	Yes	Yes	Yes
6 or more	Yes	Yes	Yes	Yes	No

[1] A range of rates is not provided because there is only a single data point.

3.3 Guiding Principles

The recommended approach for estimating trip generation for a proposed development is based on the following principles.

When the *Trip Generation* data plot contains more than 20 data points and a regression curve and equation are provided, use of the regression equation is recommended.

A regression equation with an R^2 of at least 0.75 is preferred because it indicates the desired level of correlation between the trips generated by a site and the value measured for an independent variable.

For the same reason, a weighted average rate is preferred when the standard deviation is less than or equal to 110 percent of the weighted average rate.

The value of the independent variable for the study site must fall within the range of data included to use either the rate or equation. Otherwise local data are needed.

Supplemental local data are suggested when the data plot has less than six data points.

The number of trips determined by either the rate or the equation should fall within the cluster of data points (i.e., the range of trip values) found at the study site's independent variable value. Otherwise, additional local data are needed.

Use Regression Equation When:
◆ Regression equation is provided
◆ Independent variable is within range of data *and*
◆ Either the data plot has at least 20 points or $R^2 \geq 0.75$, equation falls within data cluster in plot, and standard deviation > 110 percent of weighted average rate

Use Weighted Average Rate When:
◆ At least three data points
◆ Independent variable is within range of data
◆ Standard deviation ≤ 110 percent of weighted average rate
◆ $R^2 < 0.75$ or no equation is provided
◆ Weighted average rate falls within data cluster in plot

Collect Local Data When:
◆ Study site is not compatible with ITE land use code definition
◆ Only 1 or 2 data points; preferably when five or fewer data points
◆ Independent variable does not fall within range of data
◆ Neither weighted average rate line or fitted curve fall within data cluster at size of development

In order to put these principles into practice, two alternative approaches are available to the analyst. The highlighted box in this section presents a checklist for choosing between using the weighted average rate, using the regression equa-

tion and collecting local data. A detailed step-by-step approach for estimating trip generation is presented in the next section.

3.4 Recommended Procedure for Estimating Trip Generation

A step-by-step procedure for determining how best to estimate trip generation using data contained in *Trip Generation* is shown below. The procedure is also outlined with simplified text in the flow chart in Figure 3.1.

Step 1: Is the development under analysis consistent with the description of the land use code in *Trip Generation* and with the described or presumed characteristics of development sites for which data points are provided?

If **yes**, proceed to Step 2. If **no**, collect local data for the land use being analyzed and establish a local rate. Refer to Chapter 4 for guidelines.

Caution: The analyst should exercise caution before trying to quantify the trip generation effects of isolated and minor changes in characteristics of a particular land use. *Trip Generation* data are compiled from a wide range of sources with a potentially high variability in site characteristics within the bounds of the land use code definition. *Trip Generation* does not provide information on the secondary characteristics of the surveyed sites and therefore any analysis of the effects of changes in site characteristics is purely hypothetical and not provable at this date.

Step 2: Is the size of the development under analysis (in terms of the unit of measurement of the independent variable) within the range of the data shown in the data plot?

If **yes**, proceed to Step 3.
If **no**, collect local data and establish a local rate. Refer to Chapter 4 for guidelines.

Step 3: How many data points comprise the sample reported in *Trip Generation?*

If the number of data points is **one or two**, collect local data and establish a local rate. Refer to Chapter 4 for guidelines.
If the number of data points is **three, four, or five**, the analyst is encouraged to collect local data and establish a local rate (see Chapter 4), but can otherwise proceed to Step 4. If the number of data points is **six or more**, proceed to Step 4.

Step 4: Is a regression equation provided?

If **yes**, proceed to Step 7.
If **no**, proceed to Step 5.

Figure 3.1 Recommended Procedure for Selecting Between *Trip Generation* Average Rates and Equations

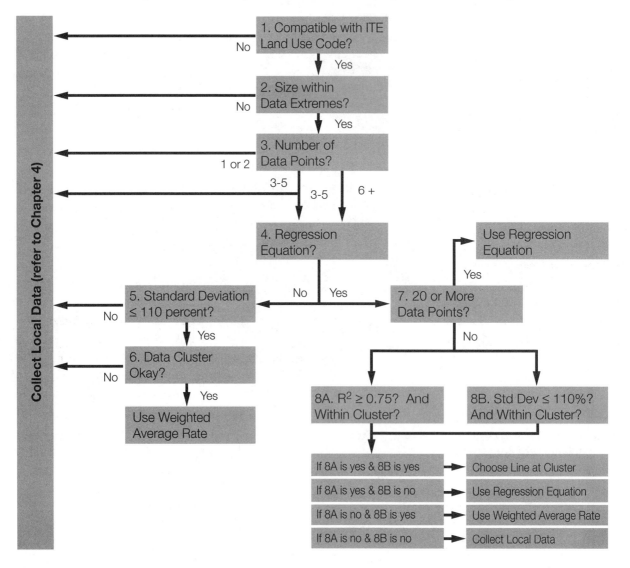

Step 5: Is the standard deviation for the weighted average rate less than or equal to 110 percent of the weighted average rate (calculation: the standard deviation divided by weighted average rate is less than or equal to 1.1)?

> If **yes**, proceed to Step 6.
> If **no**, collect local data and establish a local rate. Refer to Chapter 4 for guidelines.

Step 6: Is the line that corresponds to the weighted average rate within the cluster of data points near the size of the development site?

(Note: If there are no data points near the site size, but there are good matches at somewhat smaller and larger sizes, assume the answer is yes.)

> If **yes**, **USE THE WEIGHTED AVERAGE RATE.**
> If **no**, collect local data and establish a local rate. Refer to Chapter 4 for guidelines.

Step 7: Are at least 20 data points distributed over the range of values typically found for the independent variable? Are there few erratic data points (i.e., outliers)? Is the line corresponding to the regression equation within the cluster of data points at the size of the development in question?

If all answers are **yes**, **USE THE REGRESSION EQUATION.**

If at least one answer is **no**, proceed to Step 8.

Caution: The regression equation typically yields a line with a y-intercept. For an independent variable with a low value (i.e., near zero), the regression equation might produce a trip ends estimate that is illogical. In such a case, the analyst should use the weighted average trip rate to estimate trip ends.

Step 8: Answer Questions 8A and 8B.

Question 8A:
> Is the R^2 for the regression equation greater than or equal to 0.75? And, is the line corresponding to the regression equation within the cluster of data points at the size of the development in question?

Question 8B:
> Is the standard deviation for the weighted average rate less than or equal to 110 percent of the weighted average rate? And, is the line corresponding to the weighted average rate within the cluster of data points at the size of the development in question?

(Note: If there are no data points near the site size, but there are good matches at somewhat smaller and larger sizes, assume the answer is yes.)

If Questions 8A and 8B are both answered yes, **choose whichever line** (representing either the equation or the weighted average rate) **best fits** the data points at the size of the independent variable in question. The decision could be different for different points in the chart.

If the answer to Question 8A is yes and to Question 8B is no, **USE THE REGRESSION EQUATION.**

If the answer to Question 8A is no and to Question 8B is yes, **USE THE WEIGHTED AVERAGE RATE.**

If the answers to Questions 8A and 8B are both no, **COLLECT LOCAL DATA.** Refer to Chapter 4 for guidance. Also, if the answers to 8A and 8B are no, an acceptable **EXCEPTION** to the "collect local data" recommendation is if the rate or equation line passes through the cluster of data at the size of the development in question. If such is the case, use either the weighted average rate or the regression equation (whichever line is appropriate).

3.5 Sample Problems

The recommended step-by-step procedure for estimating trip generation is illustrated by the following sample problems. Reference is made in the problems to the data presented in *Trip Generation*, Seventh Edition. The problems are presented in the same order as their respective land use types appear in *Trip Generation*.

Problem 1: Estimate trip generation for Land Use Code 130, Industrial Park on a weekday during the morning peak hour of adjacent street traffic as a function of gross floor area (see page 143 of *Trip Generation*, Seventh Edition). Assume the site will have 800,000 sq. ft. of GFA.

> **Step 2:** size of site is within the range of data
> **Step 3:** sufficient number of data points (i.e., 40)
> **Step 4:** regression equation provided
> **Step 7:** more than 20 data points
> **Use Regression Equation**

Problem 2: Estimate trip generation for Land Use Code 223, Mid-Rise Apartment on a weekday during the morning peak hour of the generator as a function of dwelling units (see page 362). For this example, assume there will be 160 dwelling units.

> **Step 2:** size of site is within the range of data

Step 3: sufficient number of data points (seven)
Step 4: regression equation provided
Step 7: less than 20 data points
Step 8A: R^2 of 0.91 is greater than or equal to 0.75
Step 8B: standard deviation is not less than or equal to 110 percent of the weighted average rate (170 percent)
Use Regression Equation

Problem 3: Estimate trip generation for Land Use Code 252, Senior Adult Housing—Attached on a weekday during the a.m. peak hour of the generator as a function of occupied dwelling units (page 464). For this example, assume 150 occupied dwelling units.

> **Step 2:** size of site is within the range of data
> **Step 3:** four data points; could collect local data, but decide to try to use database
> **Step 4:** no regression equation provided
> **Step 5:** standard deviation is not less than or equal to 110 percent of the weighted average rate (450 percent)
> **Collect Local Data**

Problem 4: Estimate trip generation for Land Use Code 253, Congregate Care Facility on a weekday during the p.m. peak hour of the generator as a function of dwelling units (page 471).

> **Step 3:** only two data points
> **Collect Local Data**

Problem 5: Estimate trip generation for Land Use Code 310, Hotel on a weekday during the morning peak hour of adjacent street traffic as a function of occupied rooms (page 543). For this example, assume the hotel will have 300 occupied rooms.

> **Step 2:** size of site is within the range of data
> **Step 3:** sufficient number of data points (17)
> **Step 4:** regression equation provided
> **Step 7:** less than 20 data points
> **Step 8A:** R^2 of 0.69 is less than 0.75
> **Step 8B:** standard deviation is not less than or equal to 110 percent of the average rate (125 percent)
> **Step 8 Exception:** both the rate and equation lines pass through the data cluster at a size of 300 rooms
> **Use either the Regression Equation or the Weighted Average Rate**

Problem 6: Estimate trip generation for Land Use Code 444, Movie Theater with Matinee on a Sunday during the peak hour of the generator as a function of movie screens (page 776). For this example, assume the site will have only one screen.

> **Step 2:** characteristics of site are outside limits of independent variable (i.e., one screen exceeds the low of 6 screens provided in sample)
> **Collect Local Data**

Problem 7: Estimate trip generation for Land Use Code 530, High School on a weekday during the a.m. peak hour as a function of the number of employees (page 930). For this example, assume the school will have 200 employees.

> **Step 2:** number of employees is within the range of data
> **Step 3:** sufficient number of data points (52)
> **Step 4:** no regression equation provided
> **Step 5:** standard deviation is less than or equal to 110 percent of the weighted average weight (61 percent)
> **Step 6:** weighted average rate line is not within data cluster at site' snumber of employees.
> **Collect Local Data**

Problem 8: Same as problem 7, except assume the school will have 130 employees.

> **Step 2:** number of employees is within the range of data
> **Step 3:** sufficient number of data points (52)
> **Step 4:** no regression equation provided
> **Step 5:** standard deviation is less than or equal to 110 percent of the weighted average weight (61 percent)
> **Step 6:** weighted average rate line is within data cluster at site's number of employees
> **Use Weighted Average Rate**

Problem 9: Estimate trip generation for Land Use Code 550, University/College on a weekday during the a.m. peak hour of adjacent street traffic as a function of the number of employees (page 997). Assume the university/college will have 1,000 employees.

> **Step 2:** size of site is within the range of data
> **Step 3:** only four data points; but decide to try to use data base
> **Step 4:** regression equation provided
> **Step 7:** less than 20 data points
> **Step 8A:** R^2 of 0.64 is less than 0.75
> **Step 8B:** standard deviation is not less than or equal to 110 percent of the weighted average rate (140 percent)
> **Collect Local Data** at two sites and merge with ITE data base (as described in Chapter 4)

Problem 10: Estimate trip generation for Land Use Code 813, Free-Standing Discount Superstore on a weekday during the p.m. peak hour of generator as a function of gross floor area (page 1,332). For this example, assume the store size will be 180,000 square feet of GFA.

> **Step 2:** size of site is within the range of data
> **Step 3:** sufficient number of data points (nine)
> **Step 4:** regression equation provided
> **Step 7:** less than 20 data points
> **Step 8A:** R^2 of 0.55 is less than 0.75
> **Step 8B:** standard deviation is less than or equal to 110 percent of the weighted average rate (53 percent)
> **Use Weighted Average Rate**

Problem 11: Estimate trip generation for Land Use Code 866, Pet Supply Superstore (page 1,619).
> **Step 3:** only one data point
> **Collect Local Data**

CHAPTER 4
Conducting a Trip Generation Study

4.1 Background

A local jurisdiction may wish to conduct its own trip generation study to validate use of ITE *Trip Generation* rates or equations in its community, establish its own rates reflecting unique conditions found in that community, or establish rates for land use types not included in *Trip Generation*. A state or province may wish to investigate trip generation rates in detail for land use types of particular concern in its jurisdiction. Consultants, ITE districts, sections, or individual ITE members may want to supplement the ITE national database on trip generation.

To maintain consistency with ITE's nationally recognized database and procedures, local studies should follow procedures consistent with those described below. However, it is recognized that local jurisdictions may need to tailor the process to meet the specific needs of the community and the characteristics of the sites being studied.

To enhance the national database, ITE encourages the submittal of all new trip generation data. Sample data collection forms for reporting the information are included at the end of this chapter. These forms should be used whenever possible.

4.2 Reasons to Conduct a Trip Generation Study

The general purpose of a trip generation study is to collect and analyze data on the relationships between trip ends and site characteristics for a particular land use.

Before initiating the study, its specific purpose should be identified. The specific purpose will help the analyst target the characteristics of the sites, the data to be collected, the number of sites to survey and the analysis to be conducted.

◆ If the description of a site is **not covered by the land use classifications** presented in *Trip Generation*, the analyst should collect local data and establish a local rate.

When to Conduct a Trip Generation Study

◆ **New land use not covered by *Trip Generation***

◆ **Inadequate number of studies in *Trip Generation***

◆ **Size of site outside of range of *Trip Generation* data points**

◆ **Establish a local trip generation rate**

◆ **Validate *Trip Generation* for local application**

◆ **Supplement national database**

◆ If the site is located in a downtown setting, served by significant public transportation, or is the site of an extensive transportation demand management program, the site is **not consistent with the ITE data** and the analyst should collect local data and establish a local rate.

◆ If the size of a site **is not within the range of data points** presented in *Trip Generation* for the land use, the analyst should collect local data and establish a local rate.

◆ If the *Trip Generation* database has an **insufficient number of data points,** the analyst should collect local data and establish a local rate.

◆ If the *Trip Generation* database produces curves with **unsatisfactory standard deviation or regression coefficients**, the analyst should collect local data and establish a local rate.

◆ If local circumstances (e.g., age of residents, worker shifts, other differences in independent variables) make a site **noticeably different from the sites for which data were collected and reported** in *Trip Generation*, the analyst should collect local data and establish a local rate.

◆ If the site is a **multi-use development,** the analyst should refer to Chapter 7 in this handbook for guidance on special data collection and analysis efforts required for multi-use developments.

♦ If the applicability or validity of ITE *Trip Generation* data for local use is **questioned by traffic professionals or local officials,** the analyst may need to collect local data and either validate the national data or establish a local rate.

♦ If it is desirable to establish trip generation characteristics for a **land use not included in the current edition of *Trip Generation,*** the analyst should collect and analyze the data for local use and submit the data to ITE.

4.3 Trip Generation Study Design

Trip generation study design should include the land use to be surveyed, number of survey sites, selection of appropriate sites, survey period, independent variable data to be compiled and traffic counting methodology.

Information is often available from analyses undertaken in either the same jurisdiction or other jurisdictions. In planning the local study, reviewing existing data is helpful in determining issues that may be encountered and identifying expected results. Also, because existing data may be integrated into the local study to reduce the amount of new data that need to be collected, it is important to have prior knowledge of the availability and procedures used to collect the data.

Selection of Land Use to Study

Trip generation studies should be considered to supplement the *Trip Generation* database when the following conditions apply:

♦ Land uses of local interest for which ITE *Trip Generation* presents little or no data;

♦ Local land uses that do not fit into existing ITE land use classifications;

♦ Land uses that are more specific than the generalized land use categories in *Trip Generation;*

♦ Land uses for which the range in development size in the *Trip Generation* data plots does not cover the local range in development sizes; or

♦ Land uses for which local trip generation rates are theorized to be substantially different from those in *Trip Generation.*

Sample Size Determination

Sufficient sample size is necessary to enable the analyst to draw valid conclusions from the trip generation study. However, no simple statistical methodology has been established for determining the number of sites that should be studied to obtain statistically significant trip generation results. In reality, trip generation is influenced by far more than one or two

independent variables. As a result, significant variation of individual sites from the weighted average rate or regression curve is frequent. Common practice in the traffic planning industry has been to collect trip generation data at three to five sites that truly meet the recommended site selection criteria with the assumption that these data will yield a relatively stable sample.

To establish a local trip generation rate
♦ Survey at least three sites (preferably five)

To validate the ITE Trip Generation rate
♦ Survey at least three sites

To combine local trip generation data with ITE Trip Generation data
♦ Survey at least two sites

To submit data to ITE
♦ Survey at least one site

If the analyst intends to establish a local trip generation rate, it is recommended that at least three sites (and preferably at least five) be surveyed. The higher number is suggested because it will enable the analyst to more readily identify—and potentially discard—outlier values and to produce a more reliable estimate of local trip generation characteristics. It is recognized, however, that budgetary constraints and perhaps even the lack of suitable survey sites may limit the trip generation study to three sites.

If the analyst intends to validate *Trip Generation* data for local use, it is recommended that no fewer than three sites be surveyed. If the analyst intends to supplement the *Trip Generation* data with local data and produce a consolidated trip generation rate, it is recommended that at least two sites be counted. ITE will accept data from one site.

Site Selection

Site selection is critical in achieving representative and consistent trip generation rates. Failure to select sites appropriately may lead to inaccurate trip generation rates and equations. Use of unrepresentative sites as a basis for trip generation estimates can result in over- or underestimating trips to be generated by a proposed development.

Suggested criteria for identifying sites for collection of trip generation data are as follows:

◆ Data should be transferrable; therefore, it is critical that both trip data and development characteristics be representative of the land uses to be analyzed. This includes development size, mix of development components and geographic location with respect to the area roadway network and area development patterns.

◆ The development should have reasonably full occupancy (i.e., at least 85 percent) and appear to be economically healthy (note: percent

occupancy at the time of the survey, if applicable, should be recorded).

◆ The development should be mature (i.e., at least two years old) and located in a mature area so it represents the ultimate characteristics of a "successful" development.

◆ The data needed to describe the independent variables should be available.

◆ The site should be selected on the ability to obtain accurate trip generation and development characteristics.

◆ It should be possible to isolate the site for counting purposes:

- No shared parking (unless the parking areas for the site are easily distinguishable);
- No shared driveways (unless the driveways for the site are easily distinguishable);
- Limited ability for pedestrians to walk into the site from nearby parcels;
- Limited transit availability or use (unless transit usage can be counted—e.g., elementary students who ride a school bus); and
- No through-traffic.

◆ The site should have a limited number of driveways (as a data collection cost consideration).

Key Site Selection Criteria

◆ **Satisfies definition of ITE Land Use Code**

◆ **Reasonably full occupancy**

◆ **Mature**

◆ **Necessary data can be obtained readily and accurately**

◆ **Typical of sites in area with no unusual activities underway**

◆ The driveways (or the method of counting traffic) should ensure against double-counting vehicles.

◆ It should be possible for counts to be made safely. The need for any special security measures should be identified.

◆ The site should consist of a single land use activity (unless a multi-use study is being conducted as described in Chapter 7).

◆ There should be minimal to no on-site construction or adjacent roadway construction.

◆ Permission should be obtained from the owner or the building manager (note: owners/managers are sometimes more willing to be surveyed if the confidentiality of their site is guaranteed or if the results are provided to them).

Independent Variable Selection

For a new land use being surveyed, one or more appropriate independent variables need to be identified, measured and analyzed. When identifying a potential independent variable, the following points should be considered.

◆ The data for the independent variable should be readily available, for the survey site and any potential proposed development of this land use type for which trip generation estimates may be desired.

◆ The number of trips generated at the site should be influenced in a logical way by the independent variable. **Correlation does not equal causation**.

◆ Available site data should be accurate, for sites being counted and proposed future development (i.e., if it cannot be projected for new development, it is not an appropriate independent variable).

◆ Variables for similar sites should be provided directly and not merely estimated from a different variable. For example, the number of employees at a site may appear to be a valid independent variable, but it should not be used if the value is typically derived by factoring in another independent variable, such as gross square footage of the development site.

Key Characteristics of Independent Variables to Include in a Local Trip Generation Study

◆ **Logical relationship to site trip generation**

◆ **Value measured directly for the survey site, not derived**

◆ **At least those used in similar *Trip generation* land uses**

◆ **Confidence that the available site data are accurate**

When in doubt about which independent variables may be most appropriate, refer to *Trip Generation* under the same or similar land use to see which ones have produced the most stable relationships and reliable rates or equations. Typical independent variables include number of employees, gross floor area, gross leasable area and number of occupied rooms or dwelling units. It is critical that the definitions of independent variables be the same as those defined in *Trip Generation* (Chapter 3 in the *User's Guide*, Seventh Edition) or Appendix D of this handbook.

For a trip generation study involving a land use for which trip end and independent variable information is already provided in *Trip Generation*, the choice of independent variables should (at the minimum) include those presented in *Trip Generation*. If other independent variables appear to be logical and satisfy the criteria cited above, data for them should be collected and analyzed as well.

In general, it is recommended that data be collected and compiled for as many potential, appropriate independent variables as practical. As the *Trip Generation* database grows, it is quite likely that future analyses of the available data will identify additional relationships involving more than the currently used independent variables.

The sample data collection forms presented at the end of this chapter contain a list of suggested data to obtain.

Development Data Requirements

Trip generation estimates are based on development characteristics that are used as independent variables. This normally requires a check with the owners or managers of the development to ensure the availability of accurate data on physical characteristics. For example, it is insufficient to merely count dwelling units or square feet. A count of *occupied* square feet or units is needed. The occupied space represents the portion of the development that is actually generating trips.

Occupied square footage should also be carefully evaluated to make sure that it is actually being occupied *and used* rather than merely leased or purchased. For example, in some land use classifications (particularly warehousing, industrial and office), it is common practice for tenants to lease or purchase future expansion space, but not to occupy it for some time. Use of "leased or purchased" square footage instead of "occupied and used" square footage can be misleading and may be one reason for the scatter in the historical data points within certain classifications in *Trip Generation*.

Typical Development Data

◆ **Description of site**
  • **Square footage and/ or units**
  • **Percent occupancy**
  • **Site acreage**
  • **Location within area (CBD, suburban, rural)**
  • **Name and description of principal uses**

◆ **Site plan**

◆ **Adjacent street traffic volumes**

At this stage in the study design process, it is necessary to decide whether to include consideration of transit use, parking accumulation and automobile occupancy. If these issues are to be considered as factors in the analysis of the local trip generation data, then appropriate data should be collected.

Survey Periods

Site-generated traffic should be counted, if feasible, for a full 7-day period to determine when total site-generated traffic volumes peak during weekdays and the weekend. At the minimum, automatic traffic recorder counts should be taken through a full 24-hour period, although a preferred length of time would consist of 48 consecutive hours.

Some sites require manual counting techniques because automatic traffic recorder devices will not capture all trips (or may not be accurate due to the configuration of the site driveways). Manual traffic counts should last for a minimum of 2 hours for each peak period, depending on whether the adjacent street traffic peak or the generator's peak is being surveyed. If the desired traffic analysis requires other periods, counts for those periods should also be obtained.

The day of the week and time of day are also important considerations in obtaining meaningful results. The purpose of the study will dictate the critical time period for analysis.

In many cases, the season of the year is also important. In general, traffic generation for land uses with little or no seasonal variation should be counted on average days. For land uses with significant seasonal variation, time periods representing the 30th to 50th highest hours of the year may be used. Retail centers and recreational uses are typical examples of land uses with significant seasonal variation.

Care should be taken to avoid making counts during special events, holidays, construction periods, bad weather, or other times when conditions at the study site or in its vicinity may affect site trip generation. The time period being surveyed should represent typical activity for the site (e.g., no data collec-

tion should be conducted during a super sale at a retail site) unless the study is specifically designed for collecting data during a peak time (e.g., holiday shopping season for retail sites).

4.4 Conducting the Study

The following guidelines should be reviewed and followed to the extent possible when collecting traffic volume and site characteristics data.

◆ Count directional traffic volumes (entering and exiting) by 15-minute period.

◆ Where directional counts cannot be made automatically, manual counts should be made during the street peak periods, plus the peak period of the generator to record the peak-hour entering and exiting volumes. Two or more days of peak period traffic counts are desirable.

◆ If possible, collect hourly traffic volume data (or obtain from the governing jurisdiction) on all streets adjacent to and with access to or from the site so that adjacent street peak hours can be determined. The traffic counts of multiple driveway volumes must be done concurrently.

◆ Surveys or traffic counts conducted on public streets may require a courtesy call to the proper governing authority. Providing a copy of the traffic volume data or the final study to either the public or private agency involved is another good policy.

> Use of ITE's data collection forms (found at the end of this chapter) is recommended. These forms identify the information needed for a successful study, and their use will increase standardization of data collection and facilitate the inclusion of the data into ITE's existing database.

◆ Data concerning the site should be obtained through interviews with the site owner or manager and, if necessary, by means of measurements.

◆ Verify automatic counts with manual counts for short period(s). If pneumatic road tube counters are used, exercise extra caution and verification because the accuracy of this equipment may degrade at low-speed traffic conditions.

◆ If the site could be considered a multi-use development, refer to Chapter 7 (Multi-Use Developments) for guidance on additional data collection needs.

◆ If pass-by data are being collected, specific intercept surveys will be needed (as described in Chapter 5, Pass-By, Primary and Diverted Linked Trips).

◆ If needed, record hourly entering and exiting traffic by vehicle classification and vehicle occupancy and compare with corresponding automatic counts to determine a factor for adjusting the raw automatic counts. This may require classifying vehicles by number of axles if automatic counters have been used. Refer to the latest edition of the ITE *Manual of Transportation Engineering Studies* for guidance.

4.5 Establishment of a Local Trip Generation Rate or Equation

This section provides guidance if the **purpose of the trip generation study is to establish a new local trip generation rate or equation.** If the purpose of the study is to validate the use of *Trip Generation* for a local application, Section 4.6 provides appropriate guidance.

It is recommended that the first analysis step in the establishment of a local trip generation rate or equation be the development of a hypothesis for why the *Trip Generation* data might not be appropriate for local application. For example, the rationale could involve the age of residents, metropolitan area characteristics and/or the availability of transit. It is important that the local community have **a common-sense rationale for believing that the local rate is or should be significantly different from that presented in *Trip Generation*.**

(Note: the absence of any data covering a particular land use is also a valid reason for conducting a local trip generation study.)

The second analysis step in the establishment of a local trip generation rate or equation should be confirmation that a local trip generation rate/equation is indeed justified. This confirmation should be predicated on satisfying the following three criteria:

◆ At least three local sites are counted **(five sites are preferable);**

◆ The weighted average rate for the counted sites is at least 15 percent higher or lower than the comparative *Trip Generation* rate **or** if *Trip Generation* provides only two or fewer data points; and

◆ The local counts provide consistent results.

If establishment of a local trip generation rate or equation is justified based on these three criteria, the next step should involve the selection of either the computed trip generation rates or equations (if applicable) as the local trip generation estimator. The development of the local rate or equation should likewise **satisfy the standards assigned to *Trip Generation* data for its use.** In other words, the local data should be used with the same caution and desire for statistical integrity as the ITE database.

As described in Chapter 3 (Guidelines for Estimating Trip

Generation), an acceptable use of
the weighted average trip generation rate requires at least three data
points with a computed **standard
deviation that is no more than
110 percent of the weighted
average rate.** The acceptable use
of a regression equation requires at
least four data points with a computed **R² of at least 0.75.**

The local trip generation documentation should clearly state the local
rates and/or equations, the situations
in which they are applicable and
what to do in situations where they
are not applicable. The documentation should also present the site-
specific information.

Consideration should be given to
submitting the data to ITE for use

in subsequent editions of *Trip
Generation*. Sources will be cited,
but the identity of specific sites will
be kept confidential. Data should
be transmitted to:

Institute of Transportation
Engineers
1099 14th St., NW, Suite 300W
Washington, D.C. 20005-3438
Tel: +1 202-289-0222
Fax: +1 202-289-7722

4.6 Validation of *Trip Generation* Rates/Equations for Local Use

This section provides guidance if
the **purpose of the trip generation study is to validate the use
of *Trip Generation* data for a
local application.** If the purpose of
the study is to establish a new local
rate, Section 4.5 provides appropriate guidance.

> **Validation of *Trip Generation*
> data for local use does not pre-
> clude the development of local
> rates with the local data.**

Validation of *Trip Generation* rates
or equations for use in a particular
locale should be accomplished
using a two-step process. The first
step is to collect local trip generation data at no fewer than three
local sites (or supplemental data
obtained from other sources to create a database of three or more
local data sites).

The second step involves analysis of
the local data and comparison of it
to the ITE *Trip Generation* data. A
Trip Generation rate/equation should
be considered valid for local use if it
meets the following criteria:

◆ The trip generation rate for each
of the locally surveyed sites falls
within one standard deviation of
the *Trip Generation* rate;

◆ Of the sites surveyed locally, at
least one has a rate higher than the
Trip Generation weighted average
rate or equation and one has a
lower rate; **or** all of the survey sites
generated trips with totals within
15 percent of the *Trip Generation*
average rate or equation (calculated
as follows: the difference between
the survey site rate and the *Trip
Generation* rate, divided by the *Trip
Generation* rate);

◆ The locally collected data generally fall within the scatter of points
shown in the current *Trip
Generation* data plot; and

◆ Common sense derived from the
local trip generation study indicates
that the *Trip Generation* data are
valid for local application.

If the local data do not meet all of
the above criteria, development of a
local rate or equation should be
considered (refer to Section 4.5 for
guidance).

Consideration should be given to
submitting the data to ITE for use
in subsequent editions of *Trip*

Generation. Sources will be cited, but the identity of specific sites will be kept confidential.

4.7 Combining *Trip Generation* and Local Data

If the *Trip Generation* database for a particular land use is relatively small (e.g., nine or fewer sites), the local community should **consider merging the national and local databases to create a consolidated trip generation rate.** It is recommended that this merging of the data sets take place **if the local and national average rates are reasonably close (e.g., within 15 percent of each other).** The merging of the two databases under those circumstances should improve the statistical strength of the overall database for the particular land use.

If the local and national rates are substantially different, refer to Section 4.5 for guidance on the establishment of a local rate.

The following procedure demonstrates the proper steps for merging the local and national databases.

This procedure can be used for any land use, time period and independent variable for which weighted average trip rates, average size of the independent variables and number of studies are provided in *Trip Generation.*

Note: This method of combining data sets does not allow precise calculation of the standard deviation or of a revised regression equation because of the unavailability of the exact data points in the ITE national database.

The basic equation for calculating a combined weighted average trip generation rate is:

(1) combined weighted average trip rate

$$= \frac{\sum \text{trip ends (ITE)} + \sum \text{trip ends (local)}}{\sum \text{independent variable units (ITE)} + \sum \text{independent variable units (local)}}$$

The parameters "$\sum$ trip ends (ITE)" and "$\sum$ independent variable units (ITE)" can be calculated from statistics provided in *Trip Generation*.

(2) $\sum$ trip ends (ITE) = (weighted average trip rate) x (average value of independent variable unit) x (number of studies)

(3) $\sum$ independent variable units (ITE) = (average value of independent variable unit) x (number of studies)

The parameters "$\sum$ independent variable units (local)" and "$\sum$ trip ends (local)" should be available from the local data base.

The following is a sample application of this process. Assume the following information is known about three local Free-Standing Discount Superstore (Land Use 813):

	Average Weekday Trip Ends	GFA (1,000 sq. ft.)
Site 1	10,000	160
Site 2	7,000	190
Site 3	9,000	135
Total	26,000	485

From page 1,328 of Volume 3, *Trip Generation*, Seventh Edition:

    weighted average trip rate = 49.21
    average value for the independent variable unit = 160
    number of studies = 10

The weighted average trip rate for the three new sites is 53.61, which is 9 percent higher than the *Trip Generation* rate. Because the new data are within 15 percent, the following calculations can be used.

From equation (2):	$\sum$ trip ends (ITE) = 46.21 x 160 x 10 = 78,736
From equation (3):	$\sum$ independent variable units (ITE) = 160 x 10 = 1,600
From local data:	$\sum$ independent variable units (local) = 485
From local data:	$\sum$ trip ends (local) = 26,000
Applying equation (1):	combined weighted average trip rate
	= (78,736 + 26,000)/(1,600 + 485) ≈ 50.2

The updated weighted average rate is 50.2 weekday trips per 1,000 sq. ft. of gross floor area.

ite≡ Institute of Transportation Engineers

Trip Generation Data Form (Part 1)

Land Use/Building Type:[1]	ITE Land Use Code:
Source:	Source No. (*ITE use only*):
Name of Development:	Day of the Week:
City: State/Province: Zip/Postal Code:	Day: Month: Year:
Country:	Metropolitan Area:

1. For fast-food land use, please specify if hamburger- or nonhamburger-based.

Detailed Description of Development:

Location Within Area:

☐ (1) CBD   ☐ (3) Suburban (Non-CBD)   ☐ (5) Rural
☐ (2) Urban (Non-CBD)   ☐ (4) Suburban CBD   ☐ (6) Freeway Interchange Area (Rural)
                                                       ☐ (7) Not Given

Independent Variable: *(include data for as many as possible)*[2]

	Actual	Estimated		Actual	Estimated
(1) Employees (#)	☐	☐	(9) Parking Spaces (% occupied: ___)	☐	☐
(2) Persons (#)	☐	☐	(10) Beds (% occupied: ___)	☐	☐
(3) Total Units (#) (indicate unit: ___)	☐	☐	(11) Seats (#)	☐	☐
(4) Occupied Units (#) (indicate unit: ___)	☐	☐	(12) Servicing Positions/Vehicle Fueling	☐	☐
(5) Gross Floor Area (sq. ft.)	☐	☐	Positions		
(% of development occupied : ___)			(13) Shopping Center % Out-parcels/pads	☐	☐
(6) Net Rentable Area (sq. ft.)	☐	☐	(14) A.M. Peak Hour Volume of Adjacent Street Traffic	☐	☐
(7) Gross Leasable Area (sq. ft.)	☐	☐	(15) P.M. Peak Hour Volume of Adjacent Street Traffic	☐	☐
(% of development occupied: ___)			(16) Other ___	☐	☐
(8) Total Acres (% developed): ___	☐	☐	(17) Other ___	☐	☐

2. Definitions for several independent variables can be found in the *Trip Generation User's Guide Glossary*.

3. Please provide all pertinent information that helps to describe the subject project. If necessary, attach a detailed report.

Other Data:

Vehicle Occupancy (#):
A.M. ___   P.M. ___
Percent by Transit:
A.M. % ___   P.M. % ___   24-hour % ___
Percent by Carpool/Vanpool:
A.M. % ___   P.M. % ___   24-hour % ___

Employees by Shift:

	Start Time	End Time	Employees (#)
First Shift:	___	___	___
Second Shift:	___	___	___
Third Shift:	___	___	___

Parking Cost on Site: Hourly ___   Daily ___

Transportation Demand Management (TDM) Information:

At the time of this study, was there a TDM program (that may have impacted the trip generation characteristics of this site) underway?

☐ No
☐ Yes (If yes, please check appropriate box/boxes, describe the nature of the TDM program(s) and provide a source for any studies that may help quantify this impact. Attach additional sheets if necessary)

☐ (1) Transit Service   ☐ (5) Employer Support Measures   ☐ (9) Tolls and Congestion Pricing
☐ (2) Carpool Programs   ☐ (6) Preferential HOV Treatments   ☐ (10) Variable Work Hours/Compressed Work Weeks
☐ (3) Vanpool Programs   ☐ (7) Transit and Ridesharing Incentives   ☐ (11) Telecommuting
☐ (4) Bicycle/Pedestrian   ☐ (8) Parking Supply and Pricing   ☐ (12) Other ___
     Facilities and Site   Management
     Improvements

Please Complete Form on Other Side

ite Institute of Transportation Engineers

Trip Generation Data Form (Part 2)

(All = All Vehicles Counted, Including Trucks; Trucks = Heavy Duty Trucks and Buses)

Summary of Driveway Volumes

	Average Weekday (M-F)						Saturday						Sunday					
	Enter		Exit		Total		Enter		Exit		Total		Enter		Exit		Total	
	All	Trucks	All	Trucks	All	Trucks	All	Trucks	All	Trucks	All	Trucks	All	Trucks	All	Trucks	All	Trucks
24-Hour Volume																		
A.M. Peak Hour of Adjacent Street Traffic (7 – 9) Time: (ex.: 7:15 - 8:15):																		
P.M. Peak Hour of Adjacent Street Traffic (4 – 6) Time:																		
A.M. Peak Hour Generator[2] Time:																		
P.M. Peak Hour Generator[3] Time:																		
Peak Hour Generator[2] Time (Weekend):																		

[1] **Highest hourly volume** between **7 a.m.** and **9 a.m.** (**4 p.m.** and **6 p.m.**).

[2] **Highest hourly volume** during the a.m. or p.m. period.

[3] **Highest hourly volume** during the entire day.

Please refer to the *Trip Generation User's Guide* for full definition of terms.

Hourly Driveway Volumes- Average Weekday (M-F)

A.M. Period	Enter		Exit		Total		Mid-Day Period	Enter		Exit		Total		P.M. Period	Enter		Exit		Total	
	All	Trucks	All	Trucks	All	Trucks		All	Trucks	All	Trucks	All	Trucks		All	Trucks	All	Trucks	All	Trucks
6:00-7:00							11:00-12:00							3:00-4:00						
6:15-7:15							11:15-12:15							3:15-4:15						
6:30-7:30							11:30-12:30							3:30-4:30						
6:45-7:45							11:45-12:45							3:45-4:45						
7:00-8:00							12:00-1:00							4:00-5:00						
7:15-8:15							12:15-1:15							4:15-5:15						
7:30-8:30							12:30-1:30							4:30-5:30						
7:45-8:45							12:45-1:45							4:45-5:45						
8:00-9:00							1:00-2:00							5:00-6:00						

☐ **Check if Part 3 and/or additional information is attached.**

Survey conducted by: Name: _____

Organization: _____

Address: _____

City/State/Zip: _____

Telephone #: _____ Fax #: _____ E-mail: _____

Please return to:

Institute of Transportation Engineers
Technical Projects Division
1099 14th Street, NW, Suite 300 West
Washington, DC 20005-3438 USA
Telephone: +1 202-289-0222
FAX: +1 202-289-7722
ITE on the Web: www.ite.org

itc Institute of Transportation Engineers

Trip Generation Data Form (Part 3)

Name/Organization: _____ City/State: _____

Telephone Number: _____

Detailed Driveway Volumes: Attach this sheet to Parts 1 and 2 if you are providing additional information.

Day of the week: _____ (All — All Vehicles Counted, Including Trucks; Trucks — Heavy Duty Trucks and Buses)

A.M. Period	Enter		Exit		Total		P.M. Period	Enter		Exit		Total	
	All	Trucks	All	Trucks	All	Trucks		All	Trucks	All	Trucks	All	Trucks
12:00-12:15							12:00-12:15						
12:15-12:30							12:15-12:30						
12:30-12:45							12:30-12:45						
12:45-1:00							12:45-1:00						
1:00-1:15							1:00-1:15						
1:15-1:30							1:15-1:30						
1:30-1:45							1:30-1:45						
1:45-2:00							1:45-2:00						
2:00-2:15							2:00-2:15						
2:15-2:30							2:15-2:30						
2:30-2:45							2:30-2:45						
2:45-3:00							2:45-3:00						
3:00-3:15							3:00-3:15						
3:15-3:30							3:15-3:30						
3:30-3:45							3:30-3:45						
3:45-4:00							3:45-4:00						
4:00-4:15							4:00-4:15						
4:15-4:30							4:15-4:30						
4:30-4:45							4:30-4:45						
4:45-5:00							4:45-5:00						
5:00-5:15							5:00-5:15						
5:15-5:30							5:15-5:30						
5:30-5:45							5:30-5:45						
5:45-6:00							5:45-6:00						
6:00-6:15							6:00-6:15						
6:15-6:30							6:15-6:30						
6:30-6:45							6:30-6:45						
6:45-7:00							6:45-7:00						
7:00-7:15							7:00-7:15						
7:15-7:30							7:15-7:30						
7:30-7:45							7:30-7:45						
7:45-8:00							7:45-8:00						
8:00-8:15							8:00-8:15						
8:15-8:30							8:15-8:30						
8:30-8:45							8:30-8:45						
8:45-9:00							8:45-9:00						
9:00-9:15							9:00-9:15						
9:15-9:30							9:15-9:30						
9:30-9:45							9:30-9:45						
9:45-10:00							9:45-10:00						
10:00-10:15							10:00-10:15						
10:15-10:30							10:15-10:30						
10:30-10:45							10:30-10:45						
10:45-11:00							10:45-11:00						
11:00-11:15							11:00-11:15						
11:15-11:30							11:15-11:30						
11:30-11:45							11:30-11:45						
11:45-12:00							11:45-12:00						

CHAPTER 5

Pass-by, Primary and Diverted Linked Trips

5.1 Background

The trip generation rates and equations contained in *Trip Generation* are derived from actual measurements of traffic generated by individual sites. These rates and equations represent vehicles entering and exiting a site at its driveways. Therefore, these volumes are appropriate for determining the total traffic to be accommodated by site driveways.

> **The pass-by trip-making phenomenon, if estimated to be significant, should be recognized when examining the traffic impact of a development on the adjacent street system.**

There are instances, however, when the total number of trips generated by a site is different from the amount of new traffic **added** to the street system by the generator. For example, retail-oriented developments such as shopping centers, discount stores, restaurants, banks, service stations and convenience markets are often located adjacent to busy streets in order to attract the motorists already on the street. These sites attract a portion of their trips from traffic passing the site on the way from an origin to an ultimate destination. These retail trips **may not add new traffic** to the adjacent street system.

Trip-making can be broken down into two major categories: **pass-by trips** and **non-pass-by trips**. In some traffic impact study applications, it is necessary to further subdivide non-pass-by trips into **primary trips** and **diverted linked trips**. These trip types are illustrated in Figure 5.1 and are defined below.

> **Types of Trips Generated by a Site**
>
> - **Pass-By Trips**
> - **Non-Pass-By Trips**
>   - **Primary Trips**
>   - **Diverted Linked Trips**

Pass-by trips are made as intermediate stops *on the way* from an origin to a primary trip destination without a route diversion. Pass-by trips are attracted from traffic passing the site *on an adjacent street* or roadway that offers direct access to the generator. **Pass-by trips are not diverted from another roadway.**

Non-pass-by trips are simply all trips generated by a site that are not pass-by trips. This term is sometimes used when diverted linked trips are not tabulated separately from primary trips.

> **Pass-by trips do not involve a route diversion to enter the site driveway.**

Primary trips are trips made for the specific purpose of visiting the generator. The stop at the generator is the primary reason for the trip. The trip typically goes from origin to generator and then returns to the origin. For example, a home-to-shopping-to-home combination of trips is a primary trip set.

Diverted linked trips are trips that are attracted from the traffic volume on roadways within the vicinity of the generator but require a diversion from that roadway to another roadway to gain access to the site. These trips could travel on highways or freeways adjacent to a generator, but without access to the generator. **Diverted linked trips add traffic to streets adjacent to a site, but may not add traffic to the area's major travel routes** (see Figure 5.1). Both pass-by and diverted linked trips may be part of a multiple-stop chain of trips.

Figure 5.1 Types of Trips

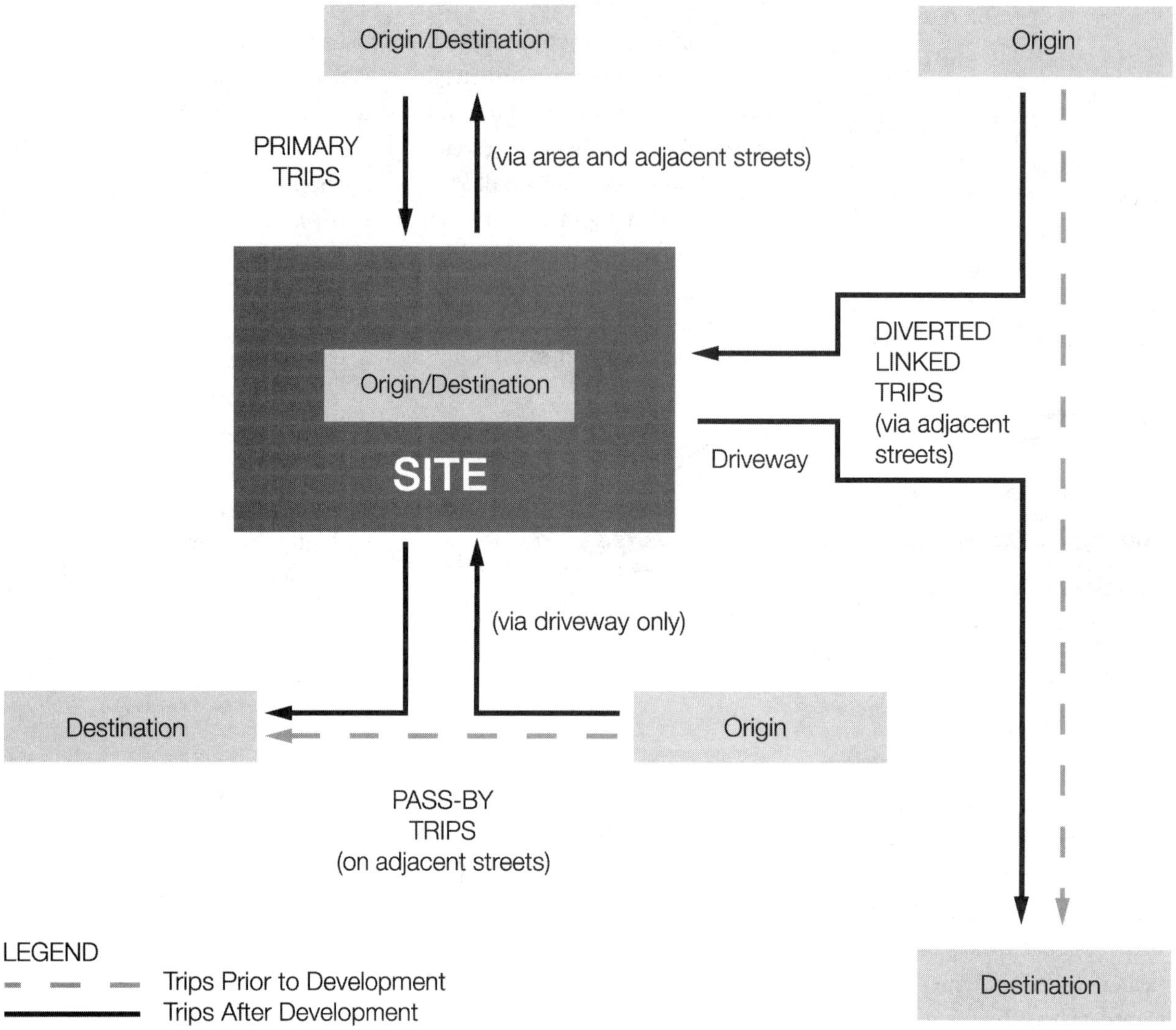
Origin/Destination
Origin
PRIMARY TRIPS
(via area and adjacent streets)
Origin/Destination
SITE
DIVERTED LINKED TRIPS (via adjacent streets)
Driveway
(via driveway only)
Destination
Origin
PASS-BY TRIPS (on adjacent streets)
LEGEND
Trips Prior to Development
Trips After Development
Destination

5.2 Sample Application of Pass-By Trip Assignment Process

The objectives in this example are to (1) estimate the number of new trips added to the adjacent street traffic volume with the development of a shopping center with 580,000 sq. ft. of gross leasable area, and (2) determine the turn movements at the shopping center driveway. The forecasted two-way evening peak hour traffic on a street adjacent to the proposed shopping center is 1,200 vehicles, as shown in Figure 5.2(A)—1,000 traveling west and 200 traveling east.

> **Objective of Assignment Process:**
>
> **Determine (1) turn movements at a shopping center driveway and (2) trips added to the adjacent street traffic volume.**

The shopping center is estimated to generate 2,000 evening peak hour trips (based on the fitted curve equation given for Land Use Code 820 on page 1,453 of *Trip Generation*, Seventh Edition). An assessment of the shopping center parking configuration and access points indicates that an estimated 20 percent of the site-generated traffic will use the driveway being analyzed in this example. Thus, the driveway volume is estimated to be 400 evening peak hour trips (i.e., 20 percent of 2,000 trips). For this

example, 50 percent enter and 50 percent exit the shopping center (as shown in Figure 5.2(B)).

From data collected at other shopping centers, it is estimated (in this example) that about 15 percent of the driveway volume is pass-by (Figure 5.2(B)). Therefore, 30 of the inbound vehicles (i.e., 15 percent of 200 vehicles) and 30 of the outbound vehicles are considered pass-by trips.

The assumed trip distribution for the non-pass-by trips is shown in Figure 5.2(C). These values are based on local knowledge of expected trip patterns for primary and diverted linked trips to and from the shopping center (based on existing travel patterns, surrounding land uses, etc.). For example, 80 percent of the non-pass-by trips are expected to arrive from the east and return to the east after the trip to the shopping center.

The distribution of pass-by trips is based on the volume of traffic passing the driveway, as shown in Figure 5.2(D). Because 83 percent of the traffic passing by the site comes from the east (i.e., 1,000 of the 1,200 shown previously in Figure 5.2 (A)), it is assumed that 83 percent of the pass-by trips will likewise arrive from the east and depart toward the west.

The assignment of the non-pass-by trips generated by the site is shown in Figure 5.2(E). The total number

of non-pass-by trips destined to the site is 170 (200 total trips minus the 30 inbound pass-by trips shown earlier in Figure 5.2(B)). Eighty percent (or 136) are expected to arrive from the east and return to the east.

The assignment of pass-by trips is shown in Figure 5.2(F). Of the 30 pass-by trips, 83 percent (or 25) arrive from the east and depart to the west. Likewise, 17 percent (or 5) arrive from the west and depart to the east. Note that the calculation also shows the expected through-trip reductions as the trips passing the site turn into the new driveway. For example, the new westbound right-turn volume of 25 causes a reduction in the westbound through movement.

The final assignment of all trips entering and leaving the shopping center driveway, as well as passing the driveway, is shown in Figure 5.2(G). These values are simply the sum of base volumes (from Figure 5.2(A)), the non-pass-by trips generated by the site (from Figure 5.2(E)), and the pass-by trips generated by the site (from Figure 5.2(F)). Note that the through-traffic volumes in both directions on the major street are reduced as a result of pass-by trip analysis.

Figure 5.2 Application of Pass-By Trips

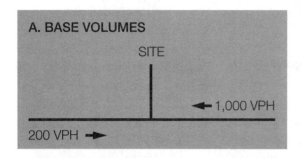

A. BASE VOLUMES

SITE

◀— 1,000 VPH

200 VPH —▶

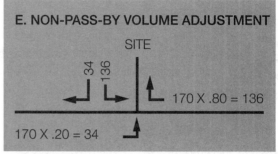

E. NON-PASS-BY VOLUME ADJUSTMENT

SITE

34 136

170 X .80 = 136

170 X .20 = 34

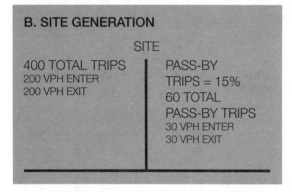

B. SITE GENERATION

SITE

400 TOTAL TRIPS
200 VPH ENTER
200 VPH EXIT

PASS-BY
TRIPS = 15%
60 TOTAL
PASS-BY TRIPS
30 VPH ENTER
30 VPH EXIT

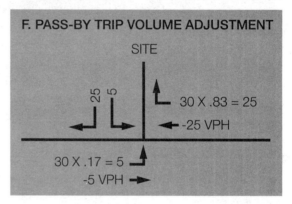

F. PASS-BY TRIP VOLUME ADJUSTMENT

SITE

25 5

30 X .83 = 25

◀— -25 VPH

30 X .17 = 5

-5 VPH —▶

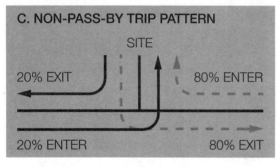

C. NON-PASS-BY TRIP PATTERN

SITE

20% EXIT   80% ENTER

20% ENTER   80% EXIT

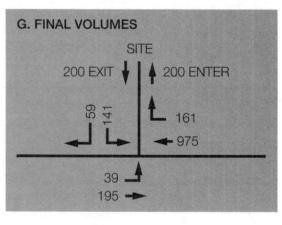

G. FINAL VOLUMES

SITE

200 EXIT   200 ENTER

59 141

161

◀— 975

39

195 —▶

LEGEND
VPH = Vehicles per hour

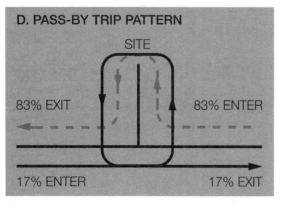

D. PASS-BY TRIP PATTERN

SITE

83% EXIT   83% ENTER

17% ENTER   17% EXIT

5.3 Cautions

Statistical analysis and correlation of pass-by data collected by the profession continue to evolve. However, due to the limited amount of pass-by data available and the inherent variability in surveyed site characteristics, it has still proven difficult to obtain high correlation indices.

> **Pass-by trips are closely linked to the size of the development and the volume of traffic on the adjacent street that can deliver the pass-by trip. However, predictive mathematical relationships have been elusive.**

Traditional pass-by trip analyses have attempted to correlate pass-by trip percentages (i.e., percentage of the total number of trips generated by a site) with units of occupied site development (such as gross leasable area, gross floor area, seats in a restaurant, or fueling positions at a gas/service station). Limited results for some land uses show that this correlation can be enhanced further by including the magnitude of the traffic passing the site on the adjacent roadways.

The analyst should exercise caution in the use of pass-by and diverted linked data presented in this chapter to ensure that the following aspects of pass-by trip characteristics are handled appropriately in the analysis process.

Diverted linked trips are clearly different from pass-by trips. Diverted linked trips add trips to the adjacent roads at a proposed or expanded site, but may not add trips to nearby major highways or freeways.

Diverted linked trips are often difficult to identify. Therefore, **diverted linked trips should be treated similarly to primary trips** unless: (1) all three (primary, pass-by and diverted linked) categories are being analyzed and processed separately, and (2) the travel routes for diverted linked trips can be clearly established.

Pass-by trips are drawn from the passing traffic stream, **but are always included in site driveway movements.** In traffic analyses, summation of driveway volumes must equal the total external site generation (i.e., the sum of primary, pass-by and diverted linked trips). Pass-by trips are not included in (and thus, subtracted from) the through-volumes passing a given site access point on an adjacent road. Standard methodologies for assessing the traffic impacts of site development typically require that diverted linked trips be included as additional trips within the confines of local impact assessment studies.

In a multi-use development, it is likely that there will be trips internal to the site (refer to Chapter 7 for guidance). Before applying the pass-by reduction, the internal trips should be removed from the total number of trips generated by the multi-use site. **Pass-by trips are only applicable to trips that enter or exit the site, not internal trips.**

Overall, diverted linked trips represent a change in local area travel patterns but constitute no new increase on a *macroscopic* scale. Within the immediate study area, diverted linked trips do represent additional traffic on individual streets and should be analyzed that way.

5.4 Database on Pass-By, Primary and Diverted Linked Trips

Listed in Table 5.1 are 22 land uses for which ITE received and compiled pass-by and diverted linked trip data. The table denotes whether the data are presented in this handbook in a table or a figure (in a data plot similar to those presented in *Trip Generation* for trip end data). Table 5.1 also identifies the time periods for which the data have been reported.

Tables 5.2 through 5.30 present the values for percentage of site generation that is accounted for by pass-by, non-pass-by, primary and diverted linked trips. For those surveys in which non-pass-by percentages are provided, no information was available on the split between primary and diverted linked trips.

Figures 5.3 through 5.16 plot the average *pass-by* trip percentages associated with the various land uses. No plots are provided for *diverted linked* trips. These figures are provided to enable the user to visualize the data scatter provided in Tables 5.2 through 5.30.

Data plots are provided for each land use where nine or more data points are available for a specific independent variable.

For all land uses except shopping centers, data are plotted for only one independent variable. For shopping centers, data are plotted for GLA and peak hour traffic on adjacent streets for the weekday evening peak period; GLA is also used as the independent variable for shopping centers during the midday Saturday time period.

A regression equation is shown on the data plot if there are more than 10 points and the R^2 is greater than 0.25 (which only occurs on two of the Land Use Code 820 data plots). Note that this threshold is less than the 0.5 threshold for R^2 used for data plots in *Trip Generation*.

Recommended guidelines for using the data presented in these figures and tables are provided in Section 5.5 of this chapter. In particular, the guidelines recommend when to use the data and how to select a pass-by percentage.

> The pass-by data in Table 5.1 were collected during *peak periods*. These pass-by relationships may differ from those during the *peak hour*.

Users of the data are cautioned that the number and geographic distribution of sites are limited. Little or no data on adjacent street traffic volumes have been collected for uses other than shopping centers. The actual pass-by and diverted linked trip percentages may vary by site due to the specific influences of the characteristics of passing traffic, area roadway network patterns, specific businesses in the site being analyzed, other nearby development and so forth. Surveys of similar developments near the analysis site are encouraged.

Because data are limited for many of the land uses, the analyst is encouraged to collect pass-by trip data and transmit the data to ITE. Section 5.6 of this chapter describes how to collect the appropriate data and provides sample forms to use.

Table 5.1 Land Uses and Time Periods with Pass-By Data

Land Use Code and Description	Time Period	Table	Figure
813 Free-Standing Discount Superstore	Weekday, p.m. Peak Period	5.2	—
815 Free-Standing Discount Store	Weekday, p.m. Peak Period	5.3	5.3
	Saturday, Midday Peak Period	5.4	5.4
816 Hardware/Paint Store	Weekday, p.m. Peak Period	5.5	—
820 Shopping Center	Weekday, p.m. Peak Period	5.6	5.5/5.6
	Saturday, Midday Peak Period	5.7	5.7
843 Automobile Parts Sales	Weekday, p.m. Peak Period	5.8	—
848 Tire Store	Weekday, p.m. Peak Period	5.9	—
850 Supermarket	Weekday, p.m. Peak Period	5.10	5.8
851 Convenience Market (Open 24 Hours)	Weekday, p.m. Peak Period	5.11	5.9
853 Convenience Market with Gasoline Pumps	Weekday, a.m. Peak Period	5.12	5.10
	Weekday, p.m. Peak Period	5.13	5.11
854 Discount Supermarket	Weekday, p.m. Peak Period	5.14	5.12
862 Home Improvement Superstore	Weekday, p.m. Peak Period	5.15	—
863 Electronics Superstore	Weekday, p.m. Peak Period	5.16	—
880 Pharmacy/Drugstore without Drive-Through Window	Weekday, p.m. Peak Period	5.17	—
881 Pharmacy/Drugstore with Drive-Through Window	Weekday, p.m. Peak Period	5.18	—
890 Furniture Store	Weekday, p.m. Peak Period	5.19	—
912 Drive-In Bank	Weekday, p.m. Peak Period	5.20	—
931 Quality Restaurant	Weekday, p.m. Peak Period	5.21	—
932 High-Turnover (Sit-Down) Restaurant	Weekday. p.m. Peak Period	5.22	5.13
934 Fast-Food Restaurant with Drive-Through Window	Weekday, a.m. Peak Period	5.23	—
	Weekday, p.m. Peak Period	5.24	5.14
935 Fast-Food Restaurant without Drive-Through WIndow and No Indoor Seating (*Specialized Land Use: Coffee/ Espresso Stand*)	Weekday	5.25/5.26	—
944 Gasoline/Service Station	Weekday, a.m. Peak Period	5.27	—
	Weekday, p.m. Peak Period	5.28	—
945 Gasoline/Service Station with Convenience Market	Weekday, a.m. Peak Period	5.29	5.15
	Weekday. p.m. Peak Period	5.30	5.16

Table 5.2
Pass-By Trips and Diverted Linked Trips
Weekday, p.m. Peak Period
Land Use 813—Free-Standing Discount Superstore

SIZE (1,000 SQ. FT. GFA)	LOCATION	WEEKDAY SURVEY DATE	NO. OF INTERVIEWS	TIME PERIOD	PRIMARY TRIP (%)	NON-PASS-BY TRIP (%)	DIVERTED LINKED TRIP (%)	PASS-BY TRIP (%)	SOURCE
146	North Olmstead, OH	Sept. 1996	210	2:45–6:45 p.m.	–	69	–	31	Mid-Ohio Regional Planning Commission
130	Ashtabula, OH	Sept. 1996	204	2:45–6:45 p.m.	–	75	–	25	Mid-Ohio Regional Planning Commission
102	Bryan, OH	Nov. 1996	100	2:45–6:45 p.m.	–	60	–	40	Mid-Ohio Regional Planning Commission
102	Oxford, OH	Oct. 1996	137	2:45–6:45 p.m.	–	72	–	28	Mid-Ohio Regional Planning Commission
218	Euclid, OH	Sept. 1996	185	2:45–6:45 p.m.	–	77	–	23	Mid-Ohio Regional Planning Commission
173	Mansfield, OH	Oct. 1996	158	2:45–6:45 p.m.	–	76	–	24	Mid-Ohio Regional Planning Commission
167	Hillsboro, OH	Oct. 1996	172	2:45–6:45 p.m.	–	70	–	30	Mid-Ohio Regional Planning Commission
167	Mentor, OH	Sept. 1996	205	2:45–6:45 p.m.	–	75	–	25	Mid-Ohio Regional Planning Commission

Average Pass-By Trip Percentage: 28

Table 5.3
Pass-By Trips and Diverted Linked Trips
Weekday, p.m. Peak Period
Land Use 815—Free-Standing Discount Store

SIZE (1,000 SQ. FT. GFA)	LOCATION	WEEKDAY SURVEY DATE	NO. OF INTERVIEWS	TIME PERIOD	PRIMARY TRIP (%)	NON-PASS-BY TRIP (%)	DIVERTED LINKED TRIP (%)	PASS-BY TRIP (%)	ADJ. STREET PEAK HOUR VOLUME	SOURCE
116	Auburn, NY	Nov. 1994	80	4:00–6:00 p.m.	33.8	—	37.4	28.8	1,490	Bergmann Associates
116	Fredonia, NY	Nov. 1994	80	4:00–6:00 p.m.	46.3	—	30.0	23.7	1,620	Bergmann Associates
122	Marlton, NJ	Nov. 1994	73	4:15–5:15 p.m.	50.7	—	27.4	21.9	1,360	Raymond Keyes Assoc.
127	Marlton, NJ	Nov. 1994	23	4:00–5:00 p.m.	21.8	—	39.1	39.1	1,410	Raymond Keyes Assoc.
127	Toms River, NJ	Nov. 1994	137	4:00–5:00 p.m.	46.0	—	40.9	13.1	1,430	Raymond Keyes Assoc.
128	Toms River, NJ	Nov. 1994	89	4:00–5:00 p.m.	60.7	—	32.6	6.7	1,290	Raymond Keyes Assoc.
128	Brick, NJ	Nov. 1994	48	4:15–5:15 p.m.	41.7	—	50.0	8.3	2,560	Raymond Keyes Assoc.
128	Brick, NJ	Nov. 1994	56	4:00–5:00 p.m.	46.4	—	39.3	14.3	2,550	Raymond Keyes Assoc.
126	Berlin, NJ	Feb. 1994	45	4:30–5:30 p.m.	75.5	—	17.8	6.7	1,230	Raymond Keyes Assoc.
126	Berlin, NJ	Feb. 1994	95	4:00–5:00 p.m.	61.0	—	37.9	1.1	1,430	Raymond Keyes Assoc.
133	Mays Landing, NJ	Feb. 1994	22	4:00–5:00 p.m.	81.8	—	9.1	9.1	3,640	Raymond Keyes Assoc.
133	Mays Landing, NJ	Feb. 1994	40	4:00–5:00 p.m.	55.0	—	42.5	2.5	3,700	Raymond Keyes Assoc.
127	Toms River, NJ	Sept. 1994	58	4:00–5:00 p.m.	65.5	—	20.7	13.8	1,380	Raymond Keyes Assoc.
127	Toms River, NJ	Sept. 1994	83	4:15–5:15 p.m.	57.8	—	28.9	13.3	1,390	Raymond Keyes Assoc.
128	Brick, NJ	Sept. 1994	117	4:30–5:30 p.m.	47.0	—	26.5	26.5	2,640	Raymond Keyes Assoc.
128	Brick, NJ	Sept. 1994	98	4:00–5:00 p.m.	49.0	—	21.4	29.6	2,640	Raymond Keyes Assoc.
127	Berlin, NJ	Sept. 1994	35	4:00–5:00 p.m.	71.4	—	20.0	8.6	1,240	Raymond Keyes Assoc.
88	Omaha, NE	n/a	n/a	4:00–6:00 p.m.	26.0	—	51.0	23.0	n/a	University of Nebraska–Lincoln
100	Omaha, NE	n/a	n/a	4:00–6:00 p.m.	32.0	—	46.0	22.0	n/a	University of Nebraska–Lincoln
100	Omaha, NE	n/a	n/a	4:00–6:00 p.m.	22.0	—	49.0	29.0	n/a	University of Nebraska–Lincoln
88	Omaha, NE	n/a	n/a	4:00–6:00 p.m.	33.0	—	48.0	19.0	n/a	University of Nebraska–Lincoln
66	Omaha, NE	n/a	n/a	4:00–6:00 p.m.	21.0	—	60.0	19.0	n/a	University of Nebraska–Lincoln

Average Pass-By Trip Percentage: 17

Figure 5.3 Free-Standing Discount Store (815)

Average Pass-By Trip Percentage vs:	**1,000 Sq. Feet Gross Floor Area**
On a:	**Weekday, p.m. Peak Period**
Number of Studies:	22
Average 1,000 Sq. Feet GFA:	118

Data Plot

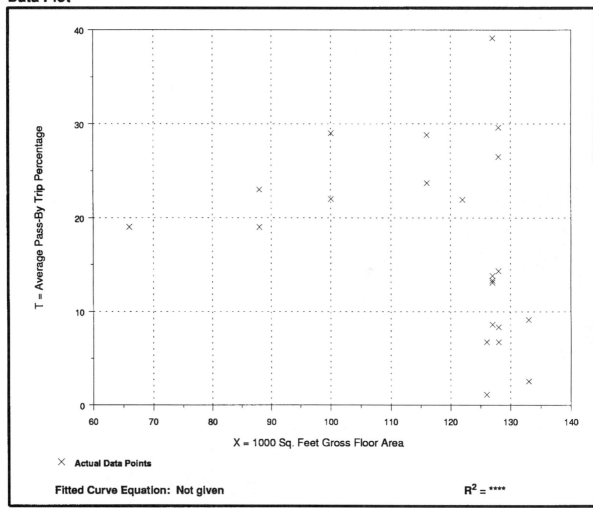

X **Actual Data Points**

Fitted Curve Equation: Not given $R^2 = $ ****

Table 5.4
Pass-By Trips and Diverted Linked Trips
Saturday, Midday Peak Period

Land Use 815—Free-Standing Discount Store

SIZE (1,000 SQ. FT. GFA)	LOCATION	SURVEY DATE	NO. OF INTERVIEWS	TIME PERIOD	PRIMARY TRIP (%)	NON-PASS-BY TRIP (%)	DIVERTED LINKED TRIP (%)	PASS-BY TRIP (%)	ADJ. STREET PEAK HOUR VOLUME	SOURCE
116	Auburn, NY	Oct. 1994	80	11:30 a.m.–12:30 p.m.	40.0	—	30.0	30.0	1,660	Bergmann Associates
116	Fredonia, NY	Nov. 1994	80	11:00 a.m.–12:00 p.m.	33.8	—	20.0	46.2	1,850	Bergmann Associates
122	Marlton, NJ	Nov. 1994	36	1:45–2:45 p.m.	41.7	—	33.3	25.0	1,810	Raymond Keyes Assoc.
127	Toms River, NJ	Nov. 1994	112	12:30–1:30 p.m.	74.1	—	10.7	15.2	1,560	Raymond Keyes Assoc.
128	Brick, NJ	Nov. 1994	61	2:00–3:00 p.m.	41.0	—	39.3	19.7	2,500	Raymond Keyes Assoc.
126	Berlin, NJ	Feb. 1994	90	12:30–1:30 p.m.	57.8	—	34.4	7.8	1,490	Raymond Keyes Assoc.
133	Mays Landing, NJ	Feb. 1994	94	1:45–2:45 p.m.	88.3	—	1.1	10.6	3,230	Raymond Keyes Assoc.
127	Toms River, NJ	Sept. 1994	131	11:30–12:30 p.m.	67.1	—	17.6	15.3	1,590	Raymond Keyes Assoc.
128	Brick, NJ	Sept. 1994	96	1:15–2:15 p.m.	43.8	—	20.8	35.4	2,640	Raymond Keyes Assoc.

Average Pass-By Trip Percentage: 23

Figure 5.4 Free-Standing Discount Store (815)

Average Pass-By Trip Percentage vs:	**1,000 Sq. Feet Gross Floor Area**
On a:	**Saturday, Midday Peak Period**
Number of Studies:	9
Average 1,000 Sq. Feet GFA:	125

Data Plot

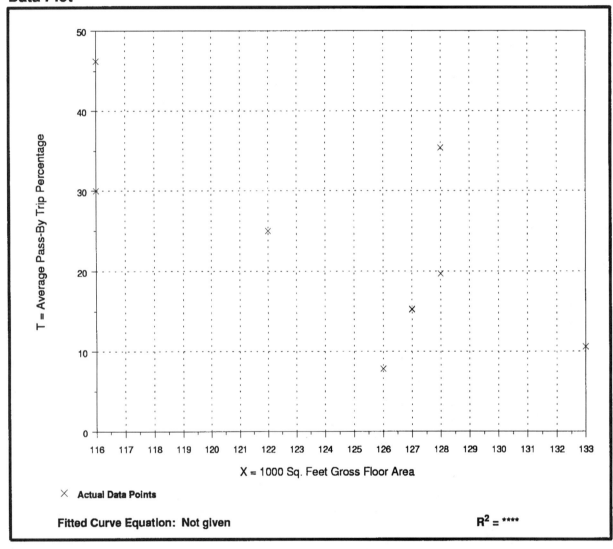

X   **Actual Data Points**

Fitted Curve Equation: Not given $R^2 = $ ****

Table 5.5
Pass-By Trips and Diverted Linked Trips
Weekday, p.m. Peak Period

Land Use 816—Hardware/Paint Store

SIZE (1,000 SQ. FT. GLA)	LOCATION	WEEKDAY SURVEY DATE	NO. OF INTERVIEWS	TIME PERIOD	PRIMARY TRIP (%)	NON-PASS-BY TRIP (%)	DIVERTED LINKED TRIP (%)	PASS-BY TRIP (%)	SOURCE
11	Aloha, OR	Nov. 1999	64	4:00–6:00 p.m.	44	—	26	30	DKS Associates
7.5	Cedar Hills, OR	Nov. 1999	33	4:00–6:00 p.m.	46	—	33	21	DKS Associates

Average Pass-By Trip Percentage: 26

Table 5.6
Pass-By Trips and Diverted Linked Trips
Weekday, p.m. Peak Period

Land Use 820 — Shopping Center

SIZE (1,000 SQ. FT. GLA)	LOCATION	WEEKDAY SURVEY DATE	NO. OF INTERVIEWS	TIME PERIOD	PRIMARY TRIP (%)	NON-PASS-BY TRIP (%)	DIVERTED LINKED TRIP (%)	PASS-BY TRIP (%)	ADJ. STREET PEAK HOUR VOLUME	AVERAGE 24-HOUR TRAFFIC	SOURCE
53	Port Orange, FL	1993	162	2:00–6:00 p.m.	—	41	—	59	n/a	n/a	TPD Inc.
9	Kissimmee, FL	1994	107	2:00–6:00 p.m.	20	—	14	66	n/a	n/a	TPD Inc.
77	Edgewater, FL	1992	365	2:00–6:00 p.m.	—	54	—	46	n/a	n/a	TPD Inc.
82	Deltona, FL	1992	336	2:00–6:00 p.m.	—	66	—	34	n/a	n/a	TPD Inc.
78	Orlando, FL	1991	702	2:00–6:00 p.m.	23	—	22	55	n/a	n/a	TPD Inc.
45	Orlando, FL	1992	844	2:00–6:00 p.m.	24	—	20	56	n/a	n/a	TPD Inc.
50	Orlando, FL	1992	555	2:00–6:00 p.m.	41	—	18	41	n/a	n/a	TPD Inc.
52	Orlando, FL	1995	665	2:00–6:00 p.m.	33	—	25	42	n/a	n/a	TPD Inc.
17	Orlando, FL	1994	196	2:00–6:00 p.m.	—	34	—	66	n/a	n/a	TPD Inc.
60	Orlando, FL	1995	1,583	3:00–7:00 p.m.	38	—	22	40	n/a	n/a	TPD Inc.
158	Crestwood, KY	Jun. 1993	129	4:00–6:00 p.m.	39	—	25	36	759	n/a	Barton-Aschman Assoc.
118	Louisville area, KY	Jun. 1993	133	4:00–6:00 p.m.	51	—	27	22	3,555	n/a	Barton-Aschman Assoc.
74	Louisville, KY	Jun. 1993	187	4:00–6:00 p.m.	43	—	27	30	922	n/a	Barton-Aschman Assoc.
59	Louisville area, KY	Jun. 1993	247	4:00–6:00 p.m.	52	—	17	31	2,659	n/a	Barton-Aschman Assoc.
145	Louisville area, KY	Jun. 1993	210	4:00–6:00 p.m.	30	—	17	53	2,636	n/a	Barton-Aschman Assoc.
104	Louisville area, KY	Jun. 1993	281	4:00–6:00 p.m.	50	—	22	28	2,111	n/a	Barton-Aschman Assoc.
235	Louisville, KY	Jun. 1993	211	4:00–6:00 p.m.	29	—	36	35	2,593	n/a	Barton-Aschman Assoc.
71	Louisville, KY	Jun. 1993	109	4:00–6:00 p.m.	42	—	33	25	1,559	n/a	Barton-Aschman Assoc.
350	Worcester, MA	Apr. 1994	224	4:00–6:00 p.m.	45	—	37	18	2,112	n/a	ICSC
738	East Brunswick, NJ	Apr. 1994	283	4:00–6:00 p.m.	79	—	7	14	8,059	n/a	ICSC
294	Philadelphia, PA	Apr. 1994	213	4:00–6:00 p.m.	51	—	24	25	4,055	n/a	ICSC
256	Hamden, CT	Apr. 1994	208	4:00–6:00 p.m.	51	—	22	27	3,422	n/a	ICSC
418	Glen Burnie, MD	Apr. 1994	281	4:00–6:00 p.m.	51	—	29	20	5,610	n/a	ICSC
560	Harrisonburg, VA	Apr. 1994	437	4:00–6:00 p.m.	49	—	32	19	3,051	n/a	ICSC

Table 5.6 (Cont'd)
Pass-By Trips and Diverted Linked Trips
Weekday, p.m. Peak Period

Land Use 820—Shopping Center

SIZE (1,000 SQ. FT. GLA)	LOCATION	WEEKDAY SURVEY DATE	NO. OF INTERVIEWS	TIME PERIOD	PRIMARY TRIP (%)	NON-PASS-BY TRIP (%)	DIVERTED LINKED TRIP (%)	PASS-BY TRIP (%)	ADJ. STREET PEAK HOUR VOLUME	AVERAGE 24-HOUR TRAFFIC	SOURCE
361	Glen Allen, VA	Apr. 1994	315	4:00–6:00 p.m.	54	—	29	17	2,034	n/a	ICSC
375	Shelby, NC	May 1994	214	4:00–6:00 p.m.	48	—	22	30	3,053	n/a	ICSC
413	Texas City, TX	May 1994	228	4:00–6:00 p.m.	52	—	20	28	589	n/a	ICSC
488	Texas City, TX	May 1994	257	4:00–6:00 p.m.	75	—	13	12	1,094	n/a	ICSC
293	Berwyn, IL	May 1994	282	4:00–6:00 p.m.	70	—	6	24	4,606	n/a	ICSC
667	Bourbonais, IL	May 1994	200	4:00–6:00 p.m.	53	—	31	16	2,770	n/a	ICSC
225	Belleville, IL	May 1994	264	4:00–6:00 p.m.	32	—	33	35	1,970	n/a	ICSC
255	Bettendorf, IA	May 1994	222	4:00–6:00 p.m.	37	—	39	24	3,706	n/a	ICSC
808	Laguna Hills, CA	Jun. 1994	240	4:00–6:00 p.m.	73	—	14	13	4,035	n/a	ICSC
450	Hanford, CA	May 1994	321	4:00–6:00 p.m.	49	—	28	23	2,787	n/a	ICSC
800	San Jose, CA	May 1994	205	4:00–6:00 p.m.	51	—	28	21	7,474	n/a	ICSC
598	Greeley, CO	May 1994	205	4:00–6:00 p.m.	55	—	28	17	3,840	n/a	ICSC
581	Pueblo, CO	May 1994	296	4:00–6:00 p.m.	53	—	29	18	2,939	n/a	ICSC
476	Bellevue, WA	May 1994	234	4:00–6:00 p.m.	54	—	20	26	3,427	n/a	ICSC
720	Framingham, MA	Dec. 1982	92	3:30–7:00 p.m.	39	—	38	23	n/a	73,628	Raymond Keyes Assoc.
890	Newark, DE	Jul. 1984	179	3:00–8:00 p.m.	49	—	39	12	n/a	n/a	Raymond Keyes Assoc.
402	Manassas, VA	Jun. 1984	87	4:00–6:00 p.m.	25	—	27	48	n/a	n/a	Raymond Keyes Assoc.
462	Ross, PA	Jun. 1980	175	5:30–7:00 p.m.	—	64	—	36	n/a	27,200	Raymond Keyes Assoc.
234	Huntington LI, NY	Nov. 1985	181	4:00–7:00 p.m.	21	—	33	46	n/a	34,630	Raymond Keyes Assoc.
658	Wayne, NJ	Sept. 1984	243	3:00–6:00 p.m.	61	—	12	27	n/a	85,600	Raymond Keyes Assoc.
1,200	Washington, DC	1980	364	4:00–6:00 p.m.	35	—	40	25	n/a	n/a	Gorove-Slade
800	Southern CA	n/a	1,000	4:00–6:00 p.m.	45	—	43	12	n/a	n/a	Frischer
451	Portland, OR	n/a	n/a	5:00–6:00 p.m.	—	75	—	25	n/a	n/a	Buttke
113	Portland, OR	n/a	n/a	5:00–6:00 p.m.	—	83	—	17	n/a	n/a	Buttke

Table 5.6 (Cont'd)
Pass-By Trips and Diverted Linked Trips
Weekday, p.m. Peak Period

Land Use 820—Shopping Center

SIZE (1,000 SQ. FT. GLA)	LOCATION	WEEKDAY SURVEY DATE	NO. OF INTERVIEWS	TIME PERIOD	PRIMARY TRIP (%)	NON-PASS-BY TRIP (%)	DIVERTED LINKED TRIP (%)	PASS-BY TRIP (%)	ADJ. STREET PEAK HOUR VOLUME	AVERAGE 24-HOUR TRAFFIC	SOURCE
622	Ramsey, MN	Nov. 1985	46	4:00–9:00 p.m.	26	—	30	44	n/a	36,370	Raymond Keyes Assoc.
736	Pensacola, FL	Oct. 1985	383	3:00–7:00 p.m.	35	—	39	26	n/a	n/a	Raymond Keyes Assoc.
84	Dover, DE	Jul. 1985	218	3:30–7:00 p.m.	6	—	44	50	n/a	n/a	Raymond Keyes Assoc.
500	Meriden, CT	Apr. 1985	n/a	4:00–6:00 p.m.	—	92	—	8	n/a	n/a	Connecticut DOT
660	Enfield, CT	Apr. 1985	n/a	4:00–6:00 p.m.	—	78	—	22	n/a	n/a	Connecticut DOT
845	Waterford, CT	Apr. 1985	n/a	4:00–6:00 p.m.	—	86	—	14	n/a	n/a	Connecticut DOT
1,060	West Hartford, CT	Apr. 1985	n/a	4:00–6:00 p.m.	—	83	—	17	n/a	n/a	Connecticut DOT
131	Pr. Georges Co., MD	1982/83	88	4:00–6:00 p.m.	—	11	—	89	n/a	n/a	JHK
181	Pr. Georges Co., MD	1982/83	105	4:00–6:00 p.m.	—	64	—	36	n/a	n/a	JHK
100	Pr. Georges Co., MD	1982/83	93	4:00–6:00 p.m.	—	64	—	36	n/a	n/a	JHK
475	Pr. Georges Co., MD	1982/83	130	4:00–6:00 p.m.	—	80	—	20	n/a	n/a	JHK
60	Pr. Georges Co., MD	1982/83	72	4:00–6:00 p.m.	—	18	—	82	n/a	n/a	JHK
90	Pr. Georges Co., MD	1982/83	91	4:00–6:00 p.m.	—	42	—	58	n/a	n/a	JHK
78	Pr. Georges Co., MD	1982/83	113	4:00–6:00 p.m.	—	41	—	59	n/a	n/a	JHK
44	Pr. Georges Co., MD	1982/83	97	4:00–6:00 p.m.	—	49	—	51	n/a	n/a	JHK
467	Pr. Georges Co., MD	1982/83	99	4:00–6:00 p.m.	—	44	—	56	n/a	n/a	JHK
352	W. Orange, NJ	Mar. 1986	149	4:00–6:00 p.m.	19	—	43	38	n/a	21,520	Raymond Keyes Assoc.
176	Tarpon Springs, FL	May 1986	124	3:00–7:00 p.m.	28	—	35	37	n/a	34,080	Raymond Keyes Assoc.
762	Orlando, FL	Fall 1985	182	4:00–6:00 p.m.	52	—	23	25	n/a	n/a	Kimley-Horn and Assoc. Inc.
166	Orlando, FL	Fall 1985	124	4:00–6:00 p.m.	48	—	25	27	n/a	n/a	Kimley-Horn and Assoc. Inc.
129	Orlando, FL	Fall 1985	116	4:00–6:00 p.m.	50	—	22	28	n/a	n/a	Kimley-Horn and Assoc. Inc.
71	Orlando, FL	Fall 1985	81	4:00–6:00 p.m.	44	—	6	50	n/a	n/a	Kimley-Horn and Assoc. Inc.

Table 5.6 (Cont'd)
Pass-By Trips and Diverted Linked Trips
Weekday, p.m. Peak Period

Land Use 820—Shopping Center

SIZE (1,000 SQ. FT. GLA)	LOCATION	WEEKDAY SURVEY DATE	NO. OF INTERVIEWS	TIME PERIOD	PRIMARY TRIP (%)	NON-PASS-BY TRIP (%)	DIVERTED LINKED TRIP (%)	PASS-BY TRIP (%)	ADJ. STREET PEAK HOUR VOLUME	AVERAGE 24-HOUR TRAFFIC	SOURCE
921	Albany, NY	Jul. & Aug. 1985	196	4:00–6:00 P.M.	42	–	35	23	n/a	60,950	Raymond Keyes Assoc.
108	Overland Park, KS	Jul. 1988	111	4:30–5:30 p.m.	61	–	13	26	n/a	34,000	n/a
118	Overland Park, KS	Aug. 1988	123	4:30–5:30 p.m.	55	–	20	25	n/a	–	n/a
256	Greece, NY	Jun. 1988	120	4:00–6:00 p.m.	62	–	–	38	n/a	23,410	Sear Brown
160	Greece, NY	Jun. 1988	78	4:00–6:00 p.m.	71	–	–	29	n/a	57,306	Sear Brown
550	Greece, NY	Jun. 1988	117	4:00–6:00 p.m.	52	–	–	48	n/a	40,763	Sear Brown
51	Boca Raton, FL	Dec. 1987	110	4:00–6:00 p.m.	34	–	33	33	n/a	42,225	Kimley-Horn and Assoc. Inc.
1,090	Ross Twp, PA	Jul. 1988	411	2:00–8:00 p.m.	56	–	10	34	n/a	51,500	Wilbur Smith and Assoc.
97	Upper Dublin Twp, PA	Winter 1988/89	n/a	4:00–6:00 p.m.	–	59	–	41	n/a	34,000	McMahon Associates
118	Tredyffrin Twp, PA	Winter 1988/89	n/a	4:00–6:00 p.m.	–	76	–	24	n/a	10,000	Booz Allen & Hamilton
122	Lawnside, NJ	Winter 1988/89	n/a	4:00–6:00 p.m.	–	63	–	37	n/a	20,000	Pennoni Associates
126	Boca Raton, FL	Winter 1988/89	n/a	4:00–6:00 p.m.	–	57	–	43	n/a	40,000	McMahon Associates
150	Willow Grove, PA	Winter 1988/89	n/a	4:00–6:00 p.m.	–	61	–	39	n/a	26,000	Booz Allen & Hamilton
153	Broward Cnty, FL	Winter 1988/89	n/a	4:00–6:00 p.m.	–	50	–	50	n/a	85,000	McMahon Associates
153	Arden, DE	Winter 1988/89	n/a	4:00–6:00 p.m.	–	70	–	30	n/a	26,000	Orth-Rodgers & Assoc. Inc.
154	Doylestown, PA	Winter 1988/89	n/a	4:00–6:00 p.m.	–	68	–	32	n/a	29,000	Orth-Rodgers & Assoc. Inc.
164	Middletown Twp, PA	Winter 1988/89	n/a	4:00–6:00 p.m.	–	67	–	33	n/a	25,000	Booz Allen & Hamilton
166	Haddon Twp, NJ	Winter 1988/89	n/a	4:00–6:00 p.m.	–	80	–	20	n/a	6,000	Pennoni Associates
205	Broward Cnty,, FL	Winter 1988/89	n/a	4:00–6:00 p.m.	–	45	–	55	n/a	62,000	McMahon Associates

Table 5.6 (Cont'd)
Pass-By Trips and Diverted Linked Trips
Weekday, p.m. Peak Period

Land Use 820—Shopping Center

SIZE (1,000 SQ. FT. GLA)	LOCATION	WEEKDAY SURVEY DATE	NO. OF INTERVIEWS	TIME PERIOD	PRIMARY TRIP (%)	NON-PASS-BY TRIP (%)	DIVERTED LINKED TRIP (%)	PASS-BY TRIP (%)	ADJ. STREET PEAK HOUR VOLUME	AVERAGE 24-HOUR TRAFFIC	SOURCE
237	W. Windsor Twp, NJ	Winter 1988/89	n/a	4:00–6:00 p.m.	—	52	—	48	n/a	46,000	Booz Allen & Hamilton
242	Willow Grove, PA	Winter 1988/89	n/a	4:00–6:00 p.m.	—	63	—	37	n/a	26,000	McMahon Associates
297	Whitehall, PA	Winter 1988/89	n/a	4:00–6:00 p.m.	—	67	—	33	n/a	26,000	Orth-Rodgers & Assoc. Inc.
360	Broward Cnty., FL	Winter 1988/89	n/a	4:00–6:00 p.m.	—	56	—	44	n/a	73,000	McMahon Associates
370	Pittsburgh, PA	Winter 1988/89	n/a	4:00–6:00 p.m.	—	81	—	19	n/a	33,000	Wilbur Smith
150	Portland, OR	n/a	519	4:00–6:00 p.m.	6	—	26	68	n/a	25,000	Kittleson and Associates
150	Portland, OR	n/a	655	4:00–6:00 p.m.	7	—	28	65	n/a	30,000	Kittleson and Associates
760	Calgary, Alberta	Oct–Dec 1987	15,436	4:00–6:00 p.m.	39	—	41	20	n/a	n/a	City of Calgary DOT
178	Bordentown, NJ	Apr. 1989	154	2:00–6:00 p.m.	—	65	—	35	n/a	37,980	Raymond Keyes Assoc.
144	Manalapan, NJ	Jul. 1990	176	3:30–6:15 p.m.	44	—	24	32	n/a	69,347	Raymond Keyes Assoc.
549	Natick, MA	Feb. 1989	n/a	4:45–5:45 p.m.	26	—	41	33	n/a	48,782	Raymond Keyes Assoc.

Average Pass-By Trip Percentage: 34

Figure 5.5 Shopping Center (820)

Average Pass-By Trip Percentage vs:	**1,000 Sq. Feet Gross Leasable Area**
On a:	**Weekday, p.m. Peak Period**
Number of Studies:	100
Average 1,000 Sq. Feet GLA:	329

Data Plot

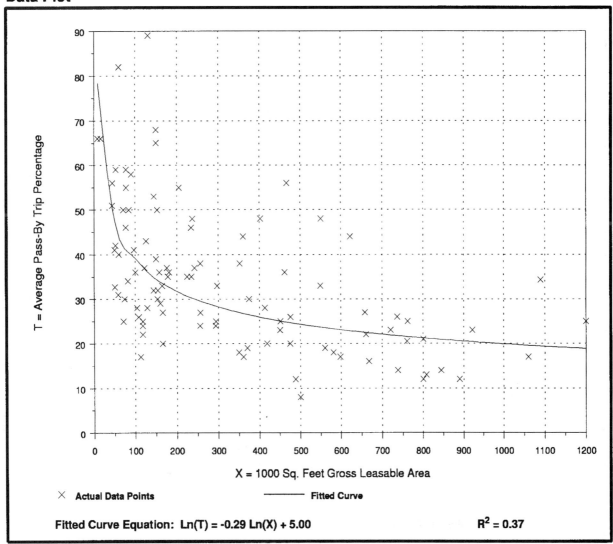

Figure 5.6 Shopping Center (820)

Average Pass-By Trip Percentage vs:	**P.M. Peak Hour Traffic on Adjacent Street**
On a:	**Weekday, p.m. Peak Period**
Number of Studies:	28
Average P.M. Peak Hr. Traf. on Adj. Street:	3,122

Data Plot

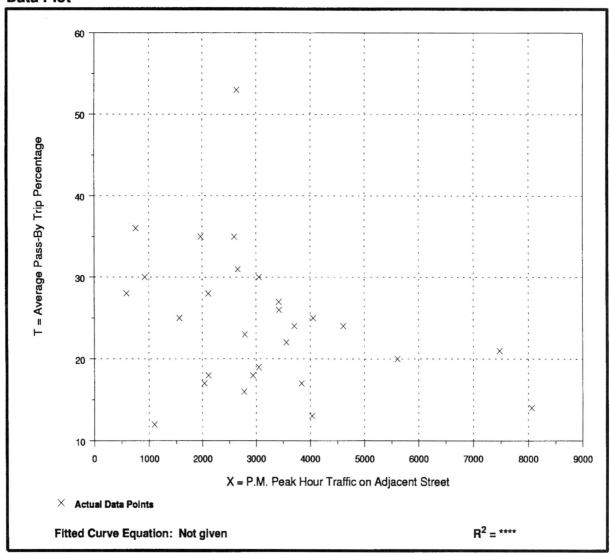

X = P.M. Peak Hour Traffic on Adjacent Street

× Actual Data Points

Fitted Curve Equation: Not given $R^2 = ****$

Table 5.7
Pass-By Trips and Diverted Linked Trips
Saturday, Midday Peak Period

Land Use 820—Shopping Center

SIZE (1,000 SQ. FT. GLA)	LOCATION	SURVEY DATE	NO. OF INTERVIEWS	TIME PERIOD	PRIMARY TRIP	NON-PASS BY TRIP (%)	DIVERTED LINKED TRIP (%)	PASS-BY TRIP (%)	AVERAGE 24-HOUR TRAFFIC	SOURCE
720	Framingham, MA	Feb. 1984	258	11:00 a.m.–4:00 p.m.	34	—	43	23	n/a	Raymond Keyes Assoc.
600	Brandywine, DE	Apr. 1983	256	10:00 a.m.–3:00 p.m.	50	—	33	17	n/a	Raymond Keyes Assoc.
880	Christiana, DE	Jul. 1984	198	11:00 a.m.–4:00 p.m.	55	—	40	5	n/a	Raymond Keyes Assoc.
234	Huntington LI, NY	Nov. 1985	223	11:00 a.m.–3:00 p.m.	22	—	39	39	n/a	Raymond Keyes Assoc.
658	Wayne, NJ	Sept. 1984	329	11:00 a.m.–4:00 p.m.	44	—	10	46	n/a	Raymond Keyes Assoc.
622	Ramsey Cnty, MN	Nov. 1985	119	11:00 a.m.–3:00 p.m.	21	—	56	23	n/a	Raymond Keyes Assoc.
736	Pensacola, FL	Oct. 1985	680	11:00 a.m.–3:00 p.m.	31	—	49	20	n/a	Raymond Keyes Assoc.
430	Ross, PA	Jun. 1980	425	11:00 a.m.–4:00 p.m.	—	78	—	22	n/a	Raymond Keyes Assoc.
176	Tampa Springs, FL	May 1986	188	11:00 a.m.–3:00 p.m.	42	—	27	31	n/a	Raymond Keyes Assoc.
144	Manalapan, NJ	Jul. 1990	264	11:00 a.m.–3:15 p.m.	47	—	22	31	63,362	Raymond Keyes Assoc.
549	Natick, MA	Feb. 1989	n/a	2:15–3:15 p.m.	39	—	33	28	48,782	Raymond Keyes Assoc.

Average Pass-By Trip Percentage: 26

Figure 5.7 Shopping Center (820)

Average Pass-By Trip Percentage vs: **1,000 Sq. Feet Gross Leasable Area**
On a: **Saturday, Midday Peak Period**
Number of Studies: 11
Average 1,000 Sq. Feet GLA: 523

Data Plot

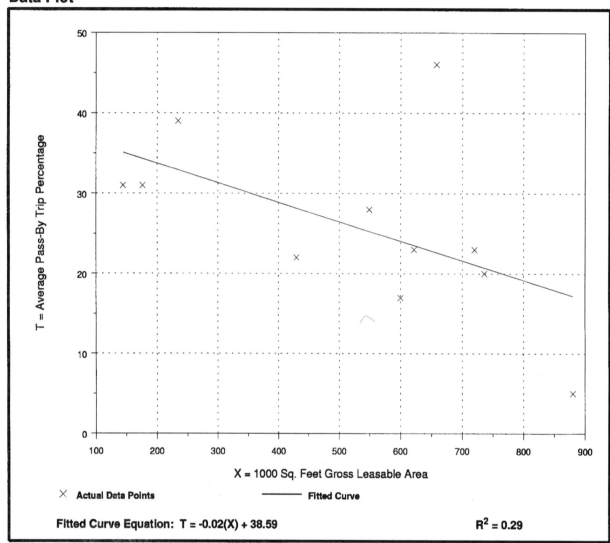

X **Actual Data Points**   ⎯⎯⎯ **Fitted Curve**

Fitted Curve Equation: T = -0.02(X) + 38.59 $R^2 = 0.29$

Table 5.8
Pass-By Trips and Diverted Linked Trips
Weekday, p.m. Peak Period

Land Use 843—Automobile Parts Sales

SIZE (1,000 SQ. FT. GFA)	LOCATION	WEEKDAY SURVEY DATE	NO. OF INTERVIEWS	TIME PERIOD	PRIMARY TRIP (%)	NON-PASS-BY TRIP (%)	DIVERTED LINKED TRIP (%)	PASS-BY TRIP (%)	ADJ. STREET PEAK HOUR VOLUME	SOURCE
15	Orlando, FL	1995	409	2:00–6:00 p.m.	44	—	13	43	n/a	TPD Inc.

Table 5.9
Pass-By Trips and Diverted Linked Trips
Weekday, p.m. Peak Period

Land Use 848—Tire Store

SIZE (1,000 SQ. FT. GFA)	LOCATION	WEEKDAY SURVEY DATE	NO. OF INTERVIEWS	TIME PERIOD	PRIMARY TRIP (%)	NON-PASS-BY TRIP (%)	DIVERTED LINKED TRIP (%)	PASS-BY TRIP (%)	ADJ. STREET PEAK HOUR VOLUME	SOURCE
4.9	Orlando, FL	1995	178	2:00–6:00 p.m.	67	—	10	23	n/a	TPD Inc.
2.8	Land O Lakes, FL	1995	46	2:00–6:00 p.m.	—	74	—	26	n/a	TPD Inc.
4.7	Orlando, FL	1988	22	2:00–6:00 p.m.	—	64	—	36	n/a	TPD Inc.

Average Pass-By Trip Percentage: 28

Table 5.10
Pass-By Trips and Diverted Linked Trips
Weekday, p.m. Peak Period

Land Use 850—Supermarket

SIZE (1,000 SQ. FT. GFA)	LOCATION	WEEKDAY SURVEY DATE	NO. OF INTERVIEWS	TIME PERIOD	PRIMARY TRIP (%)	NON-PASS-BY TRIP (%)	DIVERTED LINKED TRIP (%)	PASS-BY TRIP (%)	AVERAGE DAILY TRAFFIC	SOURCE
30	Overland Park, KS	1987	40	4:30–5:30 p.m.	48	—	20	32	n/a	n/a
<25	Chicago suburbs, IL	1987	155	3:00–6:00 p.m.	—	44	—	56	n/a	Kenig, O'Hara, Humes, Flock
<25	Chicago suburbs, IL	1987	191	3:00–6:00 p.m.	—	43	—	57	n/a	Kenig, O'Hara, Humes, Flock
<25	Chicago suburbs, IL	1987	113	3:00–6:00 p.m.	—	44	—	56	n/a	Kenig, O'Hara, Humes, Flock
34	Omaha, NE	n/a	n/a	4:00–6:00 p.m.	29	—	27	44	15,200	University of Nebraska—Lincoln
66	Omaha, NE	n/a	n/a	4:00–6:00 p.m.	30	—	47	23	63,000	University of Nebraska—Lincoln
70	Omaha, NE	n/a	n/a	4:00–6:00 p.m.	30	—	44	26	34,300	University of Nebraska—Lincoln
31	Omaha, NE	n/a	n/a	4:00–6:00 p.m.	36	—	45	19	48,700	University of Nebraska—Lincoln
31	Omaha, NE	n/a	n/a	4:00–6:00 p.m.	40	—	32	28	23,500	University of Nebraska—Lincoln
55	Omaha, NE	n/a	n/a	4:00–6:00 p.m.	35	—	38	27	27,200	University of Nebraska—Lincoln
65	Omaha, NE	n/a	n/a	4:00–6:00 p.m.	25	—	50	25	44,700	University of Nebraska—Lincoln
31	Orlando, FL	1993	440	2:00–6:00 p.m.	—	65	—	35	n/a	TPD Inc.

Average Pass-By Trip Percentage: 36

Figure 5.8 Supermarket (850)

Average Pass-By Trip Percentage vs:	**1,000 Sq. Feet Gross Floor Area**
On a:	**Weekday, p.m. Peak Period**
Number of Studies:	9
Average 1,000 Sq. Feet GFA:	46

Data Plot

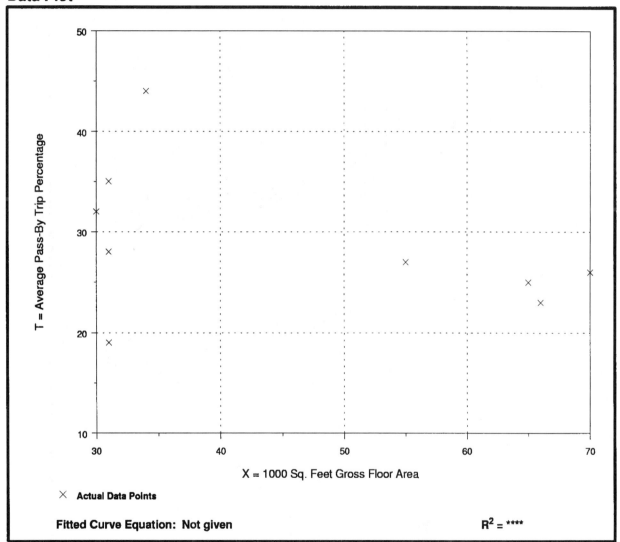

$\times$ **Actual Data Points**

Fitted Curve Equation: Not given　　　　　　　　　　　$R^2 = $ ****

Table 5.11
Pass-By Trips and Diverted Linked Trips
Weekday, p.m. Peak Period

Land Use 851—Convenience Market (Open 24 Hours)

SIZE (1,000 SQ. FT. GFA)	LOCATION	WEEKDAY SURVEY DATE	NO. OF INTERVIEWS	TIME PERIOD	PRIMARY TRIP (%)	NON-PASS-BY TRIP (%)	DIVERTED LINKED TRIP (%)	PASS-BY TRIP (%)	ADJ. STREET PEAK HOUR VOLUME	SOURCE
3	Overland Park, KS	Aug. 1987	68	4:30–5:30 p.m.	53	—	13	34	n/a	n/a
3	Overland Park, KS	Jul. 1987	68	4:30–5:30 p.m.	50	—	22	28	n/a	n/a
~1.9	Billings, MT	1987	461	4:00–6:00 p.m.	13	—	25	62	n/a	ITE Montana Section Tech Comm
<50.0	Chicago suburbs, IL	1987	72	3:00–6:00 p.m.	—	72	—	28	n/a	Kenig, O'Hara, Humes, Flock
<50.0	Chicago suburbs, IL	1987	54	3:00–6:00 p.m.	—	22	—	78	n/a	Kenig, O'Hara, Humes, Flock
<50.0	Chicago suburbs, IL	1987	34	3:00–6:00 p.m.	—	31	—	69	n/a	Kenig, O'Hara, Humes, Flock
<50.0	Chicago suburbs, IL	1987	100	3:00–6:00 p.m.	—	37	—	63	n/a	Kenig, O'Hara, Humes, Flock
<50.0	Chicago suburbs, IL	1987	43	3:00–6:00 p.m.	—	57	—	43	n/a	Kenig, O'Hara, Humes, Flock
<50.0	Chicago suburbs, IL	1987	135	3:00–6:00 p.m.	—	61	—	39	n/a	Kenig, O'Hara, Humes, Flock
<50.0	Chicago suburbs, IL	1987	74	3:00–6:00 p.m.	—	47	—	53	n/a	Kenig, O'Hara, Humes, Flock
<50.0	Chicago suburbs, IL	1987	80	3:00–6:00 p.m.	—	36	—	64	n/a	Kenig, O'Hara, Humes, Flock
2.6	Seminole Co., FL	July 1989	82	4:00–6:00 p.m.	20	—	7	73	n/a	Tipton Associates Inc.
2.6	Seminole Co., FL	July 1989	98	4:00–6:00 p.m.	15	—	4	81	n/a	Tipton Associates Inc.
2.6	Seminole Co., FL	July 1989	115	4:00–6:00 p.m.	16	—	15	69	n/a	Tipton Associates Inc.
2.6	Volusia Co., FL	July 1989	98	4:00–6:00 p.m.	15	—	11	74	n/a	Tipton Associates Inc.
2.4	Volusia Co., FL	July 1989	38	4:00–6:00 p.m.	24	—	2	74	n/a	Tipton Associates Inc.
2.6	Volusia Co., FL	July 1989	82	4:00–6:00 p.m.	8	—	5	87	n/a	Tipton Associates Inc.
2.6	Seminole Co., FL	July 1989	98	2:00–4:00 p.m.	28	—	8	64	n/a	Tipton Associates Inc.
2.4	Volusia Co., FL	July 1989	38	2:00–4:00 p.m.	21	—	11	68	n/a	Tipton Associates Inc.

Average Pass-By Trip Percentage: 61

Figure 5.9 Convenience Market (Open 24 Hours) (851)

Average Pass-By Trip Percentage vs:	**1,000 Sq. Feet Gross Floor Area**
On a:	**Weekday, p.m. Peak Period**
Number of Studies:	11
Average 1,000 Sq. Feet GFA:	2.6

Data Plot

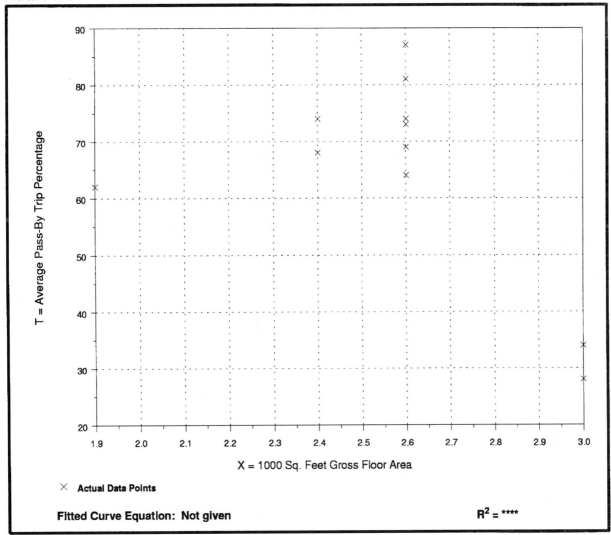

X **Actual Data Points**

Fitted Curve Equation: Not given $R^2 = {****}$

Table 5.12

Pass-By Trips and Diverted Linked Trips

Weekday, a.m. Peak Period

Land Use 853—Convenience Market with Gasoline Pumps

SIZE (1,000 SQ. FT. GFA)	LOCATION	WEEKDAY SURVEY DATE	NO. OF INTERVIEWS	TIME PERIOD	PRIMARY TRIP (%)	NON-PASS-BY TRIP (%)	DIVERTED LINKED TRIP (%)	PASS-BY TRIP (%)	ADJ. STREET PEAK HOUR VOLUME	SOURCE
2.8	Louisville area, KY	1993	n/a	7:00–9:00 a.m.	11	—	35	54	1,240	Barton-Aschman Assoc.
2.4	Louisville area, KY	1993	n/a	7:00–9:00 a.m.	17	—	35	48	1,210	Barton-Aschman Assoc.
4.2	Louisville area, KY	1993	47	7:00–9:00 a.m.	19	—	19	62	1,705	Barton-Aschman Assoc.
2.6	Crestwood, KY	1993	n/a	7:00–9:00 a.m.	15	—	13	72	940	Barton-Aschman Assoc.
3.7	Louisville area, KY	1993	49	7:00–9:00 a.m.	16	—	18	66	990	Barton-Aschman Assoc.
3.0	New Albany, IN	1993	62	7:00–9:00 a.m.	10	—	16	74	790	Barton-Aschman Assoc.
2.3	Louisville, KY	1993	58	7:00–9:00 a.m.	5	—	31	64	1,255	Barton-Aschman Assoc.
2.2	New Albany, IN	1993	79	7:00–9:00 a.m.	6	—	38	56	635	Barton-Aschman Assoc.
3.6	Louisville area, KY	1993	49	7:00–9:00 a.m.	4	—	29	67	1,985	Barton-Aschman Assoc.

Average Pass-By Trip Percentage: 63

Figure 5.10 Convenience Market with Gasoline Pumps (853)

Average Pass-By Trip Percentage vs:	**1,000 Sq. Feet Gross Floor Area**
On a:	**Weekday, a.m. Peak Period**
Number of Studies:	9
Average 1,000 Sq. Feet GFA:	3.0

Data Plot

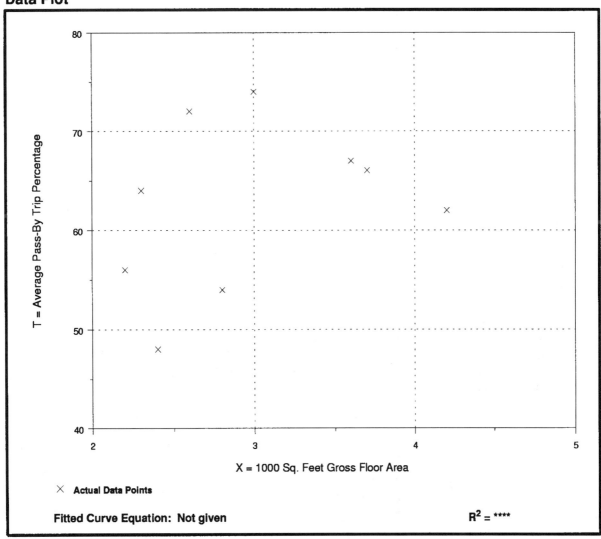

$\times$ **Actual Data Points**

Fitted Curve Equation: Not given $R^2 = $ ****

Table 5.13
Pass-By Trips and Diverted Linked Trips
Weekday, p.m. Peak Period

Land Use 853—Convenience Market with Gasoline Pumps

SIZE (1,000 SQ. FT. GFA)	LOCATION	WEEKDAY SURVEY DATE	NO. OF INTERVIEWS	TIME PERIOD	PRIMARY TRIP (%)	NON-PASS-BY TRIP (%)	DIVERTED LINKED TRIP (%)	PASS-BY TRIP (%)	ADJ. STREET PEAK HOUR VOLUME	SOURCE
2.8	Louisville area, KY	1993	n/a	4:00–6:00 p.m.	11	—	27	62	2,875	Barton-Aschman Assoc.
2.4	Louisville area, KY	1993	n/a	4:00–6:00 p.m.	13	—	29	58	2,655	Barton-Aschman Assoc.
4.2	Louisville area, KY	1993	61	4:00–6:00 p.m.	26	—	16	58	2,300	Barton-Aschman Assoc.
2.6	Crestwood, KY	1993	68	4:00–6:00 p.m.	15	—	18	67	950	Barton-Aschman Assoc.
3.7	Louisville area, KY	1993	70	4:00–6:00 p.m.	16	—	23	61	2,175	Barton-Aschman Assoc.
3.0	New Albany, IN	1993	80	4:00–6:00 p.m.	15	—	20	65	1,165	Barton-Aschman Assoc.
2.3	Louisville, KY	1993	67	4:00–6:00 p.m.	16	—	27	57	1,954	Barton-Aschman Assoc.
2.2	New Albany, IN	1993	115	4:00–6:00 p.m.	16	—	36	48	820	Barton-Aschman Assoc.
3.6	Louisville area, KY	1993	60	4:00–6:00 p.m.	17	—	27	56	2,505	Barton-Aschman Assoc.
2.6	Seminole Co., FL	1993	82	4:00–6:00 p.m.	20	—	7	73	n/a	Tipton Associates Inc.
2.6	Seminole Co., FL	1993	98	4:00–6:00 p.m.	15	—	4	81	n/a	Tipton Associates Inc.
2.6	Seminole Co., FL	1993	115	4:00–6:00 p.m.	16	—	15	69	n/a	Tipton Associates Inc.
2.6	Volusia Co., FL	1993	98	4:00–6:00 p.m.	15	—	11	74	n/a	Tipton Associates Inc.
2.4	Volusia Co., FL	1993	38	4:00–6:00 p.m.	24	—	2	74	n/a	Tipton Associates Inc.
2.7	Volusia Co., FL	1993	82	4:00–6:00 p.m.	8	—	5	87	n/a	Tipton Associates Inc.

Average Pass-By Trip Percentage: 66

Figure 5.11 Convenience Market with Gasoline Pumps (853)

Average Pass-By Trip Percentage vs:	**1,000 Sq. Feet Gross Floor Area**
On a:	**Weekday, p.m. Peak Period**
Number of Studies:	15
Average 1,000 Sq. Feet GFA:	2.8

Data Plot

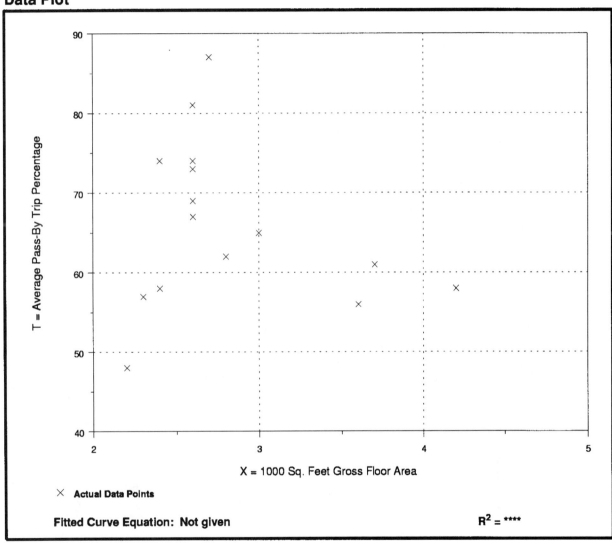

X = 1000 Sq. Feet Gross Floor Area

× **Actual Data Points**

Fitted Curve Equation: Not given $R^2 = ****$

Table 5.14
Pass-By Trips and Diverted Linked Trips
Weekday, p.m. Peak Period

Land Use 854—Discount Supermarket

SIZE (1,000 SQ. FT. GFA)	LOCATION	WEEKDAY SURVEY DATE	NO. OF INTERVIEWS	TIME PERIOD	PRIMARY TRIP (%)	NON-PASS-BY TRIP (%)	DIVERTED LINKED TRIP (%)	PASS-BY TRIP (%)	SOURCE
50	Overland Park, KS	July 1998	33	4:30–5:30 p.m.	70	—	21	9	n/a
79	Clark Cnty., WA	Nov. 2001	884	4:00–6:00 p.m.	39	—	27	34	Kittelson & Associates Inc.
80	Reno, NV	April 2002	478	4:00–6:00 p.m.	44	—	18	38	Kittelson & Associates Inc.
72	Salem, OR	Nov. 2001	827	4:00–6:00 p.m.	51	—	18	31	Kittelson & Associates Inc.
75	Hillsboro, OR	Nov. 2001	786	4:00–6:00 p.m.	40	—	27	33	Kittelson & Associates Inc.
79	Eugene, OR	Nov. 2001	637	4:00–6:00 p.m.	52	—	35	13	Kittelson & Associates Inc.
79	Yuba City, CA	April 2002	547	4:00–6:00 p.m.	64	—	21	15	Kittelson & Associates Inc.
79	Chico, CA	April 2002	798	4:00–6:00 p.m.	58	—	22	20	Kittelson & Associates Inc.
80	Antelope, CA	May 2002	617	4:00–6:00 p.m.	68	—	20	12	Kittelson & Associates Inc.
80	Elk Grove, CA	May 2002	538	4:00–6:00 p.m.	52	—	23	25	Kittelson & Associates Inc.

Average Pass-By Trip Percentage: 23

Figure 5.12 Discount Supermarket (854)

Average Pass-By Trip Percentage vs:	**1,000 Sq. Feet Gross Floor Area**
On a:	**Weekday, p.m. Peak Period**
Number of Studies:	10
Average 1,000 Sq. Feet GFA:	75

Data Plot

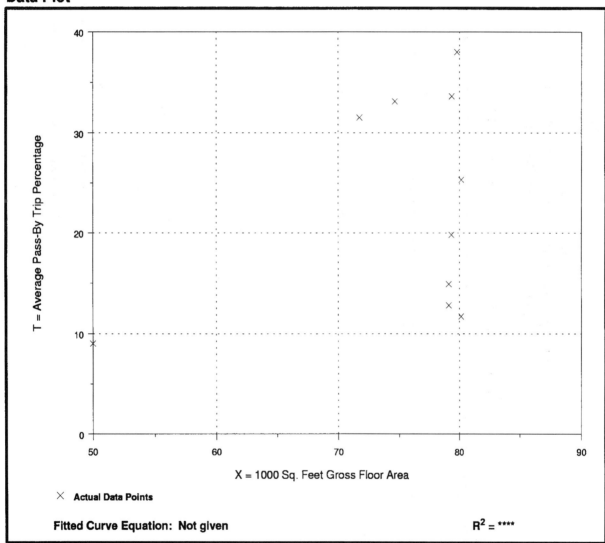

X **Actual Data Points**

Fitted Curve Equation: Not given $R^2 = ****$

Table 5.15
Pass-By Trips and Diverted Linked Trips
Weekday, p.m. Peak Period

Land Use 862—Home Improvement Superstore

SIZE (1,000 SQ. FT. GFA)	LOCATION	WEEKDAY SURVEY DATE	NO. OF INTERVIEWS	TIME PERIOD	PRIMARY TRIP (%)	NON-PASS-BY TRIP (%)	DIVERTED LINKED TRIP (%)	PASS-BY TRIP (%)	ADJ. STREET PEAK HOUR VOLUME	SOURCE
107	Casselberry, FL	1992	488	2:00–6:00 p.m.	32	—	24	44	n/a	TPD Inc.
91	Daytona Beach, FL	1993	111	2:00–6:00 p.m.	—	54	—	46	n/a	TPD Inc.
100	Orlando, FL	1993	147	2:00–6:00 p.m.	—	46	—	54	n/a	TPD Inc.

Average Pass-By Trip Percentage: 48

Table 5.16
Pass-By Trips and Diverted Linked Trips
Weekday, p.m. Peak Period

Land Use 863—Electronics Superstore

SIZE (1,000 SQ. FT. GFA)	LOCATION	WEEKDAY SURVEY DATE	NO. OF INTERVIEWS	TIME PERIOD	PRIMARY TRIP (%)	NON-PASS-BY TRIP (%)	DIVERTED LINKED TRIP (%)	PASS-BY TRIP (%)	ADJ. STREET PEAK HOUR VOLUME	SOURCE
45.7	Altamonte Springs, FL	1995	1,329	2:00–6:00 p.m.	27	—	33	40	n/a	TPD Inc.

Table 5.17
Pass-By Trips and Diverted Linked Trips
Weekday, p.m. Peak Period

Land Use 880—Pharmacy/Drugstore without Drive-Through Window

SIZE (1,000 SQ. FT. GFA)	LOCATION	WEEKDAY SURVEY DATE	NO. OF INTERVIEWS	TIME PERIOD	PRIMARY TRIP (%)	NON-PASS-BY TRIP (%)	DIVERTED LINKED TRIP (%)	PASS-BY TRIP (%)	ADJ. STREET PEAK HOUR VOLUME	SOURCE
10	Orange City, FL	1992	42	2:00–6:00 p.m.	—	35	—	65	n/a	TPD Inc.
10	Deltona, FL	1992	54	2:00–6:00 p.m.	—	40	—	60	n/a	TPD Inc.
9.6	Kissimmee, FL	1995	190	2:00–6:00 p.m.	57	—	13	30	n/a	TPD Inc.
8.6	Orlando, FL	1995	369	2:00–6:00 p.m.	25	—	15	60	n/a	TPD Inc.
13.2	New Smyrna Beach, FL	1993	55	2:00–6:00 p.m.	—	47	—	53	n/a	TPD Inc.
12	Apopka, FL	1993	365	2:00–6:00 p.m.	—	48	—	52	n/a	TPD Inc.

Average Pass-By Trip Percentage: 53

Table 5.18
Pass-By Trips and Diverted Linked Trips
Weekday, p.m. Peak Period

Land Use 881—Pharmacy/Drugstore with Drive-Through Window

SIZE (1,000 SQ. FT. GFA)	LOCATION	WEEKDAY SURVEY DATE	NO. OF INTERVIEWS	TIME PERIOD	PRIMARY TRIP (%)	NON-PASS-BY TRIP (%)	DIVERTED LINKED TRIP (%)	PASS-BY TRIP (%)	ADJ. STREET PEAK HOUR VOLUME	SOURCE
9.6	Orlando, FL	1995	370	2:00–6:00 p.m.	40	—	13	47	n/a	TPD Inc.
15.5	Orlando, FL	1995	385	2:00–6:00 p.m.	50	—	9	41	n/a	TPD Inc.
15.5	Orlando, FL	1995	522	2:00–6:00 p.m.	25	—	17	58	n/a	TPD Inc.

Average Pass-By Trip Percentage: 49

Table 5.19
Pass-By Trips and Diverted Linked Trips
Weekday, p.m. Peak Period

Land Use 890—Furniture Store

SIZE (1,000 SQ. FT. GFA)	LOCATION	WEEKDAY SURVEY DATE	NO. OF INTERVIEWS	TIME PERIOD	PRIMARY TRIP (%)	NON-PASS-BY TRIP (%)	DIVERTED LINKED TRIP (%)	PASS-BY TRIP (%)	ADJ. STREET PEAK HOUR VOLUME	SOURCE
41	Altamonte Springs, FL	1995	212	2:00–6:00 p.m.	20	—	31	49	n/a	TPD Inc.
16.5	Daytona Beach, FL	1994	39	2:00–6:00 p.m.	—	31	—	69	n/a	TPD Inc.
24	Orlando, FL	1991	103	2:00–6:00 p.m.	—	58	—	42	n/a	TPD Inc.

Average Pass-By Trip Percentage: 53

Table 5.20
Pass-By Trips and Diverted Linked Trips
Weekday, p.m. Peak Period

Land Use 912—Drive-in Bank

SIZE (1,000 SQ. FT. GFA)	LOCATION	WEEKDAY SURVEY DATE	NO. OF INTERVIEWS	TIME PERIOD	PRIMARY TRIP (%)	NON-PASS-BY TRIP (%)	DIVERTED LINKED TRIP (%)	PASS-BY TRIP (%)	ADJ. STREET PEAK HOUR VOLUME	SOURCE
16.0	Overland Park, KS	Dec. 1988	20	4:30–5:30 p.m.	55	—	30	15	n/a	n/a
3.3	Louisville area, KY	Jul. 1993	n/a	4:00–6:00 p.m.	22	—	30	48	2,570	Barton-Aschman Assoc.
3.4	Louisville area, KY	Jul. 1993	n/a	4:00–6:00 p.m.	22	—	14	64	2,266	Barton-Aschman Assoc.
3.4	Louisville area, KY	Jul. 1993	75	4:00–6:00 p.m.	11	—	32	57	1,955	Barton-Aschman Assoc.
3.5	Louisville area, KY	Jun. 1993	53	4:00–6:00 p.m.	32	—	21	47	2,785	Barton-Aschman Assoc.
6.4	Louisville area, KY	Jun. 1993	66	4:00–6:00 p.m.	20	—	27	53	2,610	Barton-Aschman Assoc.

Average Pass-By Trip Percentage: 47

Table 5.21
Pass-By Trips and Diverted Linked Trips
Weekday, p.m. Peak Period

Land Use 931—Quality Restaurant

SEATS	SIZE (1,000 SQ. FT. GFA)	LOCATION	WEEKDAY SURVEY DATE	NO. OF INTERVIEWS	TIME PERIOD	PRIMARY TRIP (%)	NON-PASS-BY TRIP (%)	DIVERTED LINKED TRIP (%)	PASS-BY TRIP (%)	ADJ. STREET PEAK HOUR VOLUME	SOURCE
240	12	Louisville area, KY	Jul. 1993	38	4:00–6:00 p.m.	36	—	38	26	4,145	Barton-Aschman Assoc.
n/a	8	Orlando, FL	1992	168	4:00–8:00 p.m.	—	55	—	45	n/a	TPD Inc.
n/a	8.8	Orlando, FL	1992	84	2:00–6:00 p.m.	40	—	16	44	n/a	TPD Inc.
n/a	6.5	Orlando, FL	1995	173	2:00–6:00 p.m.	—	38	—	62	n/a	TPD Inc.

Average Pass-By Trip Percentage: 44

Table 5.22
Pass-By Trips and Diverted Linked Trips
Weekday, p.m. Peak Period

Land Use 932—High-Turnover (Sit-Down) Restaurant

SEATS	SIZE (1,000 SQ. FT. GFA)	LOCATION	WEEKDAY SURVEY DATE	NO. OF INTERVIEWS	TIME PERIOD	PRIMARY TRIP (%)	NON-PASS-BY TRIP (%)	DIVERTED LINKED TRIP (%)	PASS-BY TRIP (%)	ADJ. STREET PEAK HOUR VOLUME	SOURCE
n/a	5.8	Orlando, FL	1992	150	2:00–6:00 p.m.	—	68	—	32	n/a	TPD Inc.
n/a	5	Casselberry, FL	1992	65	2:00–6:00 p.m.	—	42	—	58	n/a	TPD Inc.
168	5.3	Louisville area, KY	1993	24	4:00–6:00 p.m.	37	—	13	50	1,615	Barton-Aschman Assoc.
169	2.9	Louisville area, KY	1993	41	4:00–6:00 p.m.	27	—	36	37	3,935	Barton-Aschman Assoc.
150	3.1	Louisville area, KY	1993	21	4:00–6:00 p.m.	29	—	33	38	2,580	Barton-Aschman Assoc.
250	7.1	New Albany, IN	1993	n/a	4:00–6:00 p.m.	23	—	54	23	1,565	Barton-Aschman Assoc.
n/a	8	Kissimmee, FL	1995	664	2:00–6:00 p.m.	39	—	21	40	n/a	TPD Inc.
n/a	11.4	Orlando, FL	1996	267	2:00–6:00 p.m.	43	—	19	38	n/a	TPD Inc.
n/a	11.5	Orlando, FL	1996	317	2:00–6:00 p.m.	51	—	20	29	n/a	TPD Inc.
n/a	4.6	Orlando, FL	1992	276	2:00–6:00 p.m.	—	37	—	63	n/a	TPD Inc.
n/a	5.7	Orlando, FL	1994	308	2:00–6:00 p.m.	—	43	—	57	n/a	TPD Inc.
n/a	6.2	Orlando, FL	1995	521	2:00–6:00 p.m.	43	—	11	46	n/a	TPD Inc.

Average Pass-By Trip Percentage: 43

Figure 5.13 High-Turnover (Sit-Down) Restaurant (932)

Average Pass-By Trip Percentage vs:	**1,000 Sq. Feet Gross Floor Area**
On a:	**Weekday, p.m. Peak Period**
Number of Studies:	12
Average 1,000 Sq. Feet GFA:	6.4

Data Plot

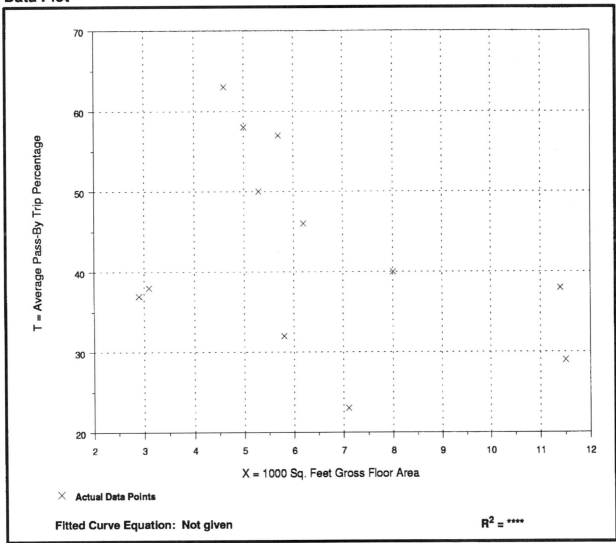

X = 1000 Sq. Feet Gross Floor Area

$\times$ **Actual Data Points**

Fitted Curve Equation: Not given $R^2 = ****$

Table 5.23
Pass-By Trips and Diverted Linked Trips
Weekday, a.m. Peak Period

Land Use 934—Fast-Food Restaurant with Drive-Through Window

SEATS	SIZE (1,000 SQ. FT. GFA)	LOCATION	WEEKDAY SURVEY DATE	NO. OF INTERVIEWS	TIME PERIOD	PRIMARY TRIP (%)	NON-PASS-BY TRIP (%)	DIVERTED LINKED TRIP (%)	PASS-BY TRIP (%)	ADJ. STREET PEAK HOUR VOLUME	SOURCE
n/a	<5	Chicago suburbs, IL	1987	84	7:00–9:00 a.m.	—	56	—	44	n/a	Kenig, O'Hara, Humes, Flock
88	1.4	Louisville area, KY	1993	n/a	7:00–9:00 a.m.	22	—	16	62	1,407	Barton-Aschman Assoc.
100	3.6	Louisville, KY	1993	n/a	7:00–9:00 a.m.	47	—	21	32	437	Barton-Aschman Assoc.
87	4.2	New Albany, IN	1993	n/a	7:00–9:00 a.m.	23	—	31	46	1,049	Barton-Aschman Assoc.
150	3.0	Louisville area, KY	1993	n/a	7:00–9:00 a.m.	14	—	43	43	2,903	Barton-Aschman Assoc.
n/a	3.3	varies	1996	n/a	6:00–9:00 a.m.	—	32	—	68	n/a	Oracle Engineering

Average Pass-By Trip Percentage: 49

Table 5.24
Pass-By Trips and Diverted Linked Trips
Weekday, p.m. Peak Period

Land Use 934—Fast-Food Restaurant with Drive-Through Window

SEATS	SIZE (1,000 SQ. FT. GFA)	LOCATION	WEEKDAY SURVEY DATE	NO. OF INTERVIEWS	TIME PERIOD	PRIMARY TRIP (%)	NON-PASS-BY TRIP (%)	DIVERTED LINKED TRIP (%)	PASS-BY TRIP (%)	ADJ. STREET PEAK HOUR VOLUME	SOURCE
n/a	~2.6	Minn-St. Paul, MN	1987	50	3:00–7:00 p.m.	27	—	48	25	n/a	n/a
n/a	<5.0	Chicago suburbs, IL	1987	80	3:00–6:00 p.m.	—	62	—	38	n/a	Kenig, O'Hara, Humes, Flock
n/a	<5.0	Chicago suburbs, IL	1987	100	3:00–6:00 p.m.	—	45	—	55	n/a	Kenig, O'Hara, Humes, Flock
n/a	<5.0	Chicago suburbs, IL	1987	159	3:00–6:00 p.m.	—	44	—	56	n/a	Kenig, O'Hara, Humes, Flock
n/a	<5.0	Chicago suburbs, IL	1987	225	3:00–6:00 p.m.	—	52	—	48	n/a	Kenig, O'Hara, Humes, Flock
n/a	<5.0	Chicago suburbs, IL	1987	88	3:00–6:00 p.m.	—	65	—	35	n/a	Kenig, O'Hara, Humes, Flock
n/a	<5.0	Chicago suburbs, IL	1987	84	3:00–6:00 p.m.	—	56	—	44	n/a	Kenig, O'Hara, Humes, Flock
88	1.3	Louisville area, KY	1993	n/a	4:00–6:00 p.m.	22	—	10	68	2,055	Barton-Aschman Assoc.
120	1.9	Louisville area, KY	1993	33	4:00–6:00 p.m.	24	—	9	67	2,447	Barton-Aschman Assoc.
87	4.2	New Albany, IN	1993	n/a	4:00–6:00 p.m.	25	—	19	56	1,632	Barton-Aschman Assoc.
150	3.0	Louisville area, KY	1993	n/a	4:00–6:00 p.m.	31	—	38	31	4,250	Barton-Aschman Assoc.
n/a	3.1	Kissimmee, FL	1995	28	2:00–6:00 p.m.	—	29	n/a	71	n/a	TPD Inc.
n/a	3.1	Apopka, FL	1996	29	2:00–6:00 p.m.	—	62	n/a	38	n/a	TPD Inc.
n/a	2.8	Winter Springs, FL	1995	47	2:00–6:00 p.m.	—	34	—	66	n/a	TPD Inc.
n/a	4.3	Longwood, FL	1994	304	2:00–6:00 p.m.	—	38	—	62	n/a	TPD Inc.
n/a	3.2	Altamonte Springs, FL	1996	202	2:00–6:00 p.m.	39	—	21	40	n/a	TPD Inc.
n/a	2.9	Winter Park, FL	1996	271	2:00–6:00 p.m.	41	—	18	41	n/a	TPD Inc.
n/a	3.3*	several	1996	varies	4:00–6:00 p.m.	—	38	—	62	n/a	Oracle Engineering

* Average of several combined studies.
Average Pass-By Trip Percentage: 50

Figure 5.14 Fast-Food Restaurant with Drive-Through Window (934)

Average Pass-By Trip Percentage vs:	**1,000 Sq. Feet Gross Floor Area**
On a:	**Weekday, p.m. Peak Period**
Number of Studies:	12
Average 1,000 Sq. Feet GFA:	3.0

Data Plot

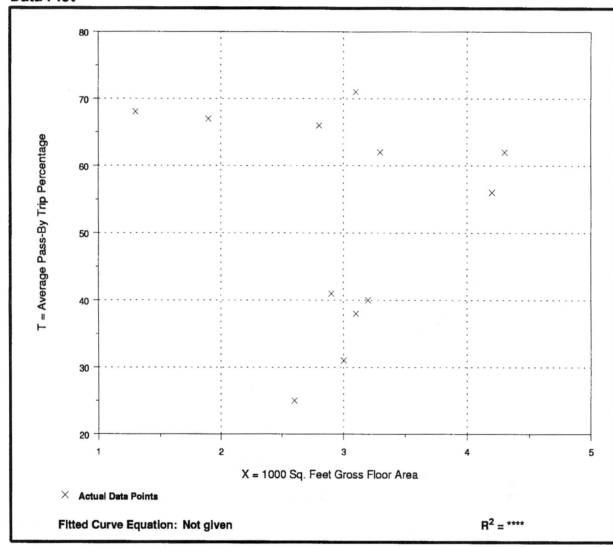

X = 1000 Sq. Feet Gross Floor Area

× **Actual Data Points**

Fitted Curve Equation: Not given　　　　　　　　$R^2 = ****$

Table 5.25
Pass-By Trips and Diverted Linked Trips
Weekday

Land Use 935—Fast-Food Restaurant with Drive-Through Window and No Indoor Seating
(Specialized Land Use: Coffee/Espresso Stand)

SIZE (1,000 SQ. FT. GFA)	LOCATION	WEEKDAY SURVEY DATE	NO. OF INTERVIEWS	TIME PERIOD	PRIMARY TRIP (%)	NON-PASS-BY TRIP (%)	DIVERTED LINKED TRIP (%)	PASS-BY TRIP (%)	SOURCE
0.1	Vancouver, WA	Nov. 1997	69	6:00 a.m.–6:00 p.m.	—	17	—	83	Kittelson & Associates Inc.

Table 5.26
Pass-By Trips and Diverted Linked Trips
Weekday

Land Use 935—Fast-Food Restaurant with Drive-Through Window and No Indoor Seating
(Specialized Land Use: Coffee/Espresso Stand)

EMPLOYEES	LOCATION	WEEKDAY SURVEY DATE	NO. OF INTERVIEWS	TIME PERIOD	PRIMARY TRIP (%)	NON-PASS-BY TRIP (%)	DIVERTED LINKED TRIP (%)	PASS-BY TRIP (%)	SOURCE
1	Vancouver, WA	Nov. 1997	70	6:00 a.m.–6:00 p.m.	—	17	—	83	Kittelson & Associates Inc.
1	Woodburn, OR	Feb. 1998	109	6:00 a.m.–6:00 p.m.	—	5	—	95	Kittelson & Associates Inc.
1	Vancouver, WA	Feb. 1998	83	6:00 a.m.–1:00 p.m.	—	11	—	89	Kittelson & Associates Inc.

Average Pass-By Trip Percentage: 89

Table 5.27
Pass-By Trips and Diverted Linked Trips
Weekday, a.m. Peak Period

Land Use 944—Gasoline/Service Station

SIZE (1,000 SQ. FT. GFA)	VEHICLE FUELING POSITIONS	LOCATION	WEEKDAY SURVEY DATE	NO. OF INTERVIEWS	TIME PERIOD	PRIMARY TRIP (%)	NON-PASS-BY TRIP (%)	DIVERTED LINKED TRIP (%)	PASS-BY TRIP (%)	ADJ. STREET PEAK HOUR VOLUME	SOURCE
2.3	6	Gaithersburg, MD	1992	37	7:00–9:00 a.m.	41	—	27	32	2,080	RBA
2.1	6	Bethesda, MD	1992	26	7:00–9:00 a.m.	23	—	19	58	2,080	RBA
1.7	6	Wheaton, MD	1992	21	7:00–9:00 a.m.	14	—	19	67	900	RBA
2.0	8	Gaithersburg, MD	1992	46	7:00–9:00 a.m.	13	—	0	87	2,235	RBA
1.2	6	Damascus, MD	1992	21	7:00–9:00 a.m.	28	—	29	43	870	RBA
0.3	12	Wheaton, MD	1992	36	7:00–9:00 a.m.	8	—	31	61	3,480	RBA

Average Pass-By Trip Percentage: 58

Table 5.28
Pass-By Trips and Diverted Linked Trips
Weekday, p.m. Peak Period

Land Use 944—Gasoline/Service Station

SIZE (1,000 SQ. FT. GFA)	VEHICLE FUELING POSITIONS	LOCATION	WEEKDAY SURVEY DATE	NO. OF INTERVIEWS	TIME PERIOD	PRIMARY TRIP (%)	NON-PASS-BY TRIP (%)	DIVERTED LINKED TRIP (%)	PASS-BY TRIP (%)	ADJ. STREET PEAK HOUR VOLUME	SOURCE
n/a	n/a	Chicago suburbs, IL	1987	48	3:00–7:00 p.m.	—	79	—	21	n/a	Kenig, O'Hara, Humes, Flock
n/a	n/a	Chicago suburbs, IL	1987	34	3:00–6:00 p.m.	—	75	—	25	n/a	Kenig, O'Hara, Humes, Flock
n/a	n/a	Chicago suburbs, IL	1987	42	3:00–6:00 p.m.	—	80	—	20	n/a	Kenig, O'Hara, Humes, Flock
2.3	6	Gaithersburg, MD	1992	55	4:00–6:00 p.m.	11	—	49	40	2,760	RBA
2.1	6	Bethesda, MD	1992	30	4:00–6:00 p.m.	20	—	27	53	1,060	RBA
1.7	6	Wheaton, MD	1992	18	4:00–6:00 p.m.	6	—	33	61	2,510	RBA
2.0	8	Gaithersburg, MD	1992	47	4:00–6:00 p.m.	23	—	15	62	2,635	RBA
1.2	6	Damascus, MD	1992	26	4:00–6:00 p.m.	11	—	31	58	1,020	RBA
0.3	12	Wheaton, MD	1992	52	4:00–6:00 p.m.	10	—	52	38	3,835	RBA

Average Pass-By Trip Percentage: 42

Table 5.29
Pass-By Trips and Diverted Linked Trips
Weekday, a.m. Peak Period

Land Use 945—Gasoline/Service Station with Convenience Market

SIZE (1,000 SQ. FT. GFA)	VEHICLE FUELING POSITIONS	LOCATION	WEEKDAY SURVEY DATE	NO. OF INTERVIEWS	TIME PERIOD	PRIMARY TRIP (%)	NON-PASS-BY TRIP (%)	DIVERTED LINKED TRIP (%)	PASS-BY TRIP (%)	ADJ. STREET PEAK HOUR VOLUME	SOURCE
0.8	8	Louisville area, KY	1993	61	7:00–9:00 a.m.	15	—	25	60	4,000	Barton-Aschman Assoc.
0.6	8	Louisville, KY	1993	48	7:00–9:00 a.m.	13	—	19	68	1,307	Barton-Aschman Assoc.
0.7	10	Louisville, KY	1993	47	7:00–9:00 a.m.	11	—	22	67	1,105	Barton-Aschman Assoc.
0.7	8	Louisville area, KY	1993	n/a	7:00–9:00 a.m.	22	—	22	56	1,211	Barton-Aschman Assoc.
0.7	10	Louisville area, KY	1993	n/a	7:00–9:00 a.m.	31	—	12	46	1,211	Barton-Aschman Assoc.
0.3	n/a	Louisville area, KY	1993	75	7:00–9:00 a.m.	15	—	13	72	n/a	Barton-Aschman Assoc.
0.8	8	Silver Spring, MD	1992	36	7:00–9:00 a.m.	14	—	39	47	3,095	RBA
0.4	8	Derwood, MD	1992	46	7:00–9:00 a.m.	0	—	25	75	3,770	RBA
2.2	8	Kensington, MD	1992	31	7:00–9:00 a.m.	34	—	19	47	1,785	RBA
1	8	Silver Spring, MD	1992	35	7:00–9:00 a.m.	9	—	13	78	7,080	RBA

Average Pass-By Trip Percentage: 62

Figure 5.15 Gasoline/Service Station with Convenience Market (945)

Average Pass-By Trip Percentage vs:	**1,000 Sq. Feet Gross Floor Area**
On a:	**Weekday, a.m. Peak Period**
Number of Studies:	10
Average 1,000 Sq. Feet GFA:	0.8

Data Plot

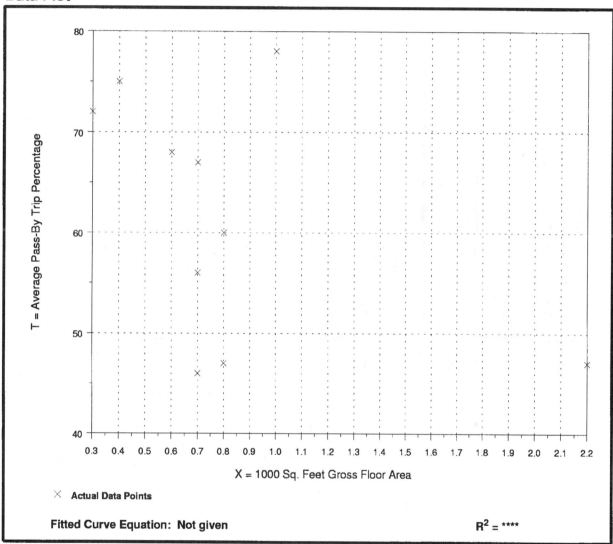

X **Actual Data Points**

Fitted Curve Equation: Not given $R^2 = ****$

Table 5.30
Pass-By Trips and Diverted Linked Trips
Weekday, p.m. Peak Period

Land Use 945—Gasoline/Service Station with Convenience Market

SIZE (1,000 SQ. FT. GFA)	VEHICLE FUELING POSITIONS	LOCATION	WEEKDAY SURVEY DATE	NO. OF INTERVIEWS	TIME PERIOD	PRIMARY TRIP (%)	NON-PASS-BY TRIP (%)	DIVERTED LINKED TRIP (%)	PASS-BY TRIP (%)	ADJ. STREET PEAK HOUR VOLUME	SOURCE
0.8	8	Louisville area, KY	1993	83	4:00–6:00 p.m.	8	—	40	52	4,965	Barton-Aschman Assoc.
0.6	8	Louisville, KY	1993	60	4:00–6:00 p.m.	20	—	27	53	1,491	Barton-Aschman Assoc.
0.7	10	Louisville, KY	1993	n/a	4:00–6:00 p.m.	19	—	24	57	1,812	Barton-Aschman Assoc.
0.7	8	Louisville area, KY	1993	n/a	4:00–6:00 p.m.	7	—	21	72	2,657	Barton-Aschman Assoc.
0.7	10	Louisville area, KY	1993	n/a	4:00–6:00 p.m.	16	—	29	55	2,657	Barton-Aschman Assoc.
0.8	8	Silver Spring, MD	1992	36	4:00–6:00 p.m.	14	—	19	67	3,095	RBA
0.4	8	Derwood, MD	1992	46	4:00–6:00 p.m.	11	—	43	46	3,770	RBA
2.1	8	Kensington, MD	1992	31	4:00–6:00 p.m.	13	—	35	52	1,785	RBA
1	8	Silver Spring, MD	1992	35	4:00–6:00 p.m.	3	—	43	54	7,080	RBA

Average Pass-By Trip Percentage: 56

Figure 5.16 Gasoline/Service Station with Convenience Market (945)

Average Pass-By Trip Percentage vs:	**1,000 Sq. Feet Gross Floor Area**
On a:	**Weekday, p.m. Peak Period**
Number of Studies:	9
Average 1,000 Sq. Feet GFA:	0.9

Data Plot

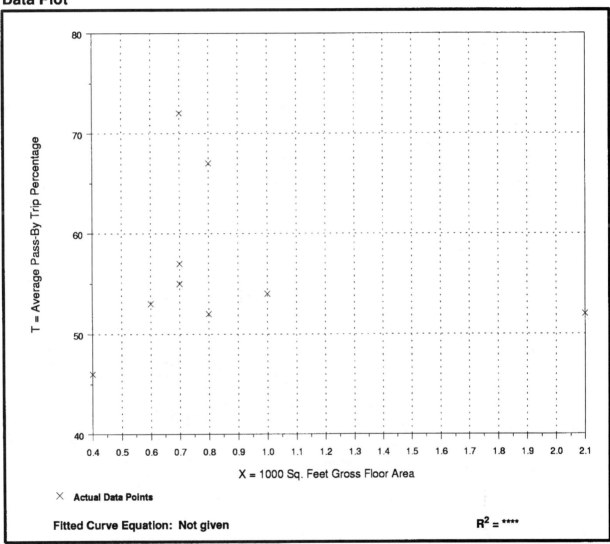

X **Actual Data Points**

Fitted Curve Equation: Not given $R^2 = ****$

5.5 A Guide for Establishing Pass-By Trip-Making Estimates

Presented below is a guide for establishing an estimate for pass-by trip-making. The analyst has two basic options: (1) derive a pass-by estimate from Tables 5.2–5.30 and Figures 5.3–5.16 presented on the preceding pages, or (2) collect relevant local data on pass-by trip-making. If the latter is chosen, surveys should be conducted at similar existing developments (on the same major roadways if feasible). Pass-by trips should be surveyed for the desired analysis hours. Adjacent street traffic volumes should also be determined for these hours. Section 5.6 provides guidance on desirable sample sizes and the survey instrument.

> **The guidance provided in this section deals directly with pass-by trip estimation. A similar procedure can be used for estimating diverted linked trips and primary trips.**

If a regression curve is provided, the equation should be used as a starting point for pass-by trip estimation. Consideration should then be given to the data scatter at the size of the independent variable in question.

If a regression equation is not provided, the average rate derived from the pass-by data presented in Section 5.4 should be considered as

> **The level of accuracy in estimating pass-by percentages is typically not as great as that available for estimating overall trip generation.**

a starting point for estimating pass-by trip-making if the following criteria are met:

• Sample consists of three or more data points; and

• Size of the proposed development (in terms of the independent variable unit of measurement) is within the range of the data points provided in the table or the figure.

The analyst should start with the average rate listed in the pertinent table and make appropriate refinements, if circumstances dictate. For example, a review of the data or of the data plot might indicate that the development site in question could be expected to have a slightly higher or lower pass-by rate due to its size, location, or proximity to through-traffic.

If the above criteria are not met for the data provided in this handbook, it is recommended that local data be collected to supplement the existing database. The analyst should also recognize that the pass-by survey results are presented in the tables and the figures regardless of the survey sample size and its effect on potential error in estimating the pass-by trips for the surveyed site.

5.6 Recommended Data Collection Procedures

Additional studies are still desired in all land use categories characterized by a significant number of pass-by trips. Transportation professionals, developers and others are encouraged to collect and submit pass-by data for these land uses. These additional studies will help in the future refinement of pass-by trip estimation procedures and will enable analysts to produce more accurate forecasts.

> **Need For Additional Data to**
>
> ◆ **Broaden database on observed pass-by percentages**
>
> ◆ **Identify and evaluate more independent variables against which to correlate pass-by percentages**
>
> ◆ **Validate the percentages shown in this handbook**
>
> ◆ **Add data for new land use classifications**

The following text describes the data needed for a pass-by study, potential independent variables, a suggested approach for collecting pass-by trip information (including a preferred survey instrument and a recommended sample size) and the data that should be submitted to ITE to refine the pass-by trip database.

Table 5.31 Minimum Sample Size for Pass-By Trip Surveys
(95% level of confidence)

MAXIMUM ERROR IN MEAN	EXPECTED PERCENT PASS-BY TRIPS						
	20%	30%	40%	50%	60%	70%	Unknown
10%	61	81	92	96	92	81	96
15%	27	36	41	43	41	36	43

Pass-By Survey Sample Size

The number of pass-by interviews should meet the minimum sample size requirements listed in Table 5.31. The analyst should make an initial estimate of the expected percentage of pass-by trips. Figures 5.3 through 5.16 and Tables 5.2 through 5.30 can be used to derive this initial estimate. For example, for a 95 percent level of confidence and maximum error of 10 percent, a shopping center would require a minimum of 96 usable interviews if the expected pass-by trip percentage is 50 percent. If the expected pass-by percentage is 20 percent, the minimum sample size is 61 usable interviews.

It is suggested that this number of interviews be conducted on all sides of the site with access in order to not favor one roadway or one area of the site over another. If different driveways are anticipated to exhibit significantly different pass-by percentages, the minimum sample size should be met for each.

Survey Instrument

A sample form for conducting the pass-by survey is presented in Figure 5.17. Interviewers should intercept persons as they exit the site. Motorists can be interviewed quickly as they walk to their vehicle, or an interview station can be set up near the parking lot exit.

A survey response of "yes" to Question 2 should be counted as a **primary trip**. A survey response of "yes" to Question 3 should be counted as a **pass-by trip**. The remainder of the trips should be considered **diverted linked trips**.

Data Submittal

The pass-by trip survey results (as well as the site trip generation data) should be summarized in a format similar to that presented in Figure 5.18. A report that presents the site-specific information and the survey results is also an invaluable reference. At the minimum, include the following information:

◆ Name, address and location (city, state/province, metropolitan area) of development (note: ITE will keep the identity of the site confidential);

◆ Land use classification (shopping center, bank, fast-food restaurant, etc.) especially if it varies from the existing ITE land use code description;

◆ ITE land use code;

◆ Survey date and day of the week;

◆ Time period during which the pass-by survey was conducted;

◆ Number of interviews conducted;

◆ Volume of traffic on adjacent streets that have full or partial access to the site (hourly volume during survey period);

◆ Size of the development measured in the same units as specified in *Trip Generation*; and

◆ Size of the development in terms of other independent variables that could be collected and forecast and that might correlate to pass-by trip-making.

Data will be accepted and sources cited in subsequent editions of *Trip Generation*. The identity of specific sites will be kept confidential. Data should be transmitted to:

Institute of Transportation Engineers
1099 14th Street, NW, Suite 300W
Washington, DC 20005-3438
Tel: +1 202-289-0222
Fax: +1 202-289-7722

Figure 5.17 Sample Questionnaire
for Pass-By Trip Interview

Name of Development _____

Location of Surveyor _____

Date of Survey _____

Sample Survey Instrument

TIME OF DAY	Q1. Where did your trip begin immediately prior to arriving at this site?	Q2. Will you go directly back to your origin from here?	Q3. Would you have driven by this site if you had not stopped here now?	Q4. If No to Q.3, how many miles out of your way did you travel to get here?
	A. Home B. Work C. Other Retail D. Other	Y. Yes [end of survey] N. No	Y. Yes [end of survey] N. No	

Figure 5.18
Pass-By Trip and Diverted Linked Trip Interview Results
Summary Form
[to be completed after data collection has been completed]

Name of Development _____

Description of Land Use _____

Potential Independent Variables

ITE Land Use Code_____ | Gross Square Feet _____

Date of Survey_____ | Gross Leasable Area _____

Day of Week _____ | # of Seats _____

City, State/Province_____ | # of Rooms(total/occpd)_____

Metropolitan Area _____ | # of Fueling Positions _____

Time Period of Survey _____ to _____ | Other (specify) _____

Number of Interviews _____

	SURVEY PERIOD	PEAK HOUR (_____ TO _____)
Total Site Volume: Inbound		
Outbound		
Pass-By Percentage:		
Usable Interviews:		
Adjacent Street Traffic[1]		

[1]Adjacent street traffic includes all traffic with direct access to a development site. In some cases where the site is served by some form of service roadway or roadways, the adjacent street or streets would be those that lead to the service roads, and thus may not actually be contiguous to the site.

CHAPTER 6
Estimating Trip Generation for Generalized Land Uses

6.1 Background

Trip Generation provides rates, equations and data plots for specific land use categories. These are appropriate for use in estimating trip generation for development proposals with specific land uses, such as may be known prior to applying for a site plan approval, special use permit, or driveway permit.

When?
If specific land use is not known

Accuracy of Procedure
Order-of-magnitude, at best

However, in early land development stages, there may be a need to develop an approximate estimate of future trip generation, even though the specific proposed land use is not yet known. In that case, the analyst knows nothing more than the (proposed) zoning classification, or sometimes just the anticipated zoning, of the land in question. In such cases, there would not be a site plan, so the precise quantity of development may not be known.

The following is a procedure that can be used in such situations. *Users of this procedure are cautioned that the resulting estimates should be considered order-of-magnitude at best, and must be recalculated and the analysis redone once a specific land use and development size are known.*

6.2 Overview of Recommended Procedure

There are three basic steps to the recommended process:

◆ Determine the potential mix (or mixes) of land uses in the site;

◆ Estimate the quantity of development (e.g., gross square footage, dwelling units, acres); and

◆ Estimate the number of trips generated by each individual land use and for the total site.

6.3 Determination of Potenial Land Use Mix

The appropriate trip generation estimate should reflect, to the extent possible, the specific uses within the known or assumed generalized (usually zoning) classification. These land uses can best be determined by knowing the specific ITE land use classifications permissible and prevailing locally within the (zoning) classification. If there is no zoning (existing or proposed) at the time of analysis, the prevailing use(s) in the area should be used.

To determine likely land uses, the analyst should first read the zoning ordinance to determine permissible uses. Once a list of likely land uses is complete, the analyst should review it with appropriate local public officials to determine the **prevailing mix of land uses** within recently developed areas

under the proposed zoning classification, preferably in similar locales and with similar site size and access characteristics.

> In a hypothetical example, a city's shopping center zoning classification may permit shopping centers, restaurants, bowling alleys, and banks. However, in the type of area being studied, all recent development has averaged about 80 percent shopping center and 20 percent restaurant by GLA and GSF, respectively (with no bowling alleys or banks).

6.4 Estimation of Development Quantity

Typical development densities (GSF/acre, DU/acre, or an appropriate similar ratio) should be obtained for the chosen land uses (e.g., shopping center and restaurant for the above example). Listings of values for typical development densities can be found in national publications. No typical values are presented here, however, for a specific reason—"typical" development densities from a national data source are unlikely to match "local typical" conditions. In fact, "typical" development densities may vary in different parts of a region. Because of the inherent imprecision of the trip estimating process presented herein, **it is critical that an effort be made to be as locally accurate as possible in terms of the land use mix and density.** Densities need to be discussed and agreed upon with the local planning or zoning department.

> **Use of the prevailing floor area ratio (FAR) for the area is recommended, rather than the maximum allowable FAR.**

Typical floor-area-ratio (FAR) values vary substantially as a function of land value and, in turn, the presence or absence of structured parking (which would enable higher FAR values). Therefore, a key decision to be made at this stage of the analysis is whether or not to assume structured parking. Again, prevailing development practices in similar circumstances should be assumed.

6.5 Estimation of Trip Generation

Use the procedure described in Chapter 3 to determine the preferred method for estimating trip generation for each land use. The procedure will direct the analyst either to use the weighted average rate, to use the regression equation, or to collect local data.

Trip generation rates should be developed in the most detailed development units possible. For example, GSF is preferable to acres.

Estimation of trips by each individual land use within the subject site will also enable the analyst to evaluate the potential impacts of different trip distribution patterns for different land uses.

The trip generation estimate in terms of total trip ends (T) is obtained by multiplying the trip generation rate (R) times the weighted density (D) times the area (A), or

$$T = (R \times D \times A)$$

Remember that the variables in the above equation must be measured in compatible units. For example, if R is expressed in "trips per GFA," then D could be expressed in "GFA/acre" and A in "acres."

The product of D and A in the above equation is simply the number of development units that match the independent variable for the assumed trip generation rate, based either on the weighted average rate or the regression equation.

The trip generation estimate (T) is obtained by summing the product of the rate, density and acres of the individual land uses, or

$$T = \sum_i (R_i \times D_i \times A_i)$$

where i corresponds to individual land uses or developments within the site.

Depending on the purpose of the study and its desired level of precision, testing a range of scenarios may be wise (i.e., with varying land use mixes and densities). Likewise, depending on the level of uncertainty regarding the land use mix and development density, a range of scenarios should be tested.

It should be emphasized again that this method can be used for general planning and land use application matters, and that it is not appropriate when undertaking detailed traffic impact analysis studies.

CHAPTER 7

Multi-Use Development

7.1 Background

A basic premise behind the data presented in *Trip Generation* is that data were collected at single-use, free-standing sites. However, the development of mixed-use or multi-use sites is increasingly popular. While the trip generation rates for individual uses on such sites may be the same or similar to what they are for free-standing sites, there is potential for interaction among those uses within the multi-use site, particularly where the trip can be made by walking. As a result, the total generation of vehicle trips entering and exiting the multi-use site may be reduced from simply a sum of the individual, discrete trips generated by each land use.

A common example of this internal trip-making occurs at a multi-use development containing offices and a shopping/service area. Some of the trips made by office workers to shops, restaurants, or banks may occur on site. These types of trips are defined as internal to (i.e., "captured" within) the multi-use site.

7.2 What Is a Multi-Use Development?

For purposes of this handbook, a *multi-use development is typically a single real-estate project that consists of two or more ITE land use classifications between which trips can be made without using the off-site road system.* Because of the nature of these land uses, the

trip-making characteristics are interrelated, and some trips are made among the on-site uses. This capture of trips internal to the site has the net effect of reducing vehicle trip generation between the overall development site and the external street system (compared to the total number of trips generated by comparable, stand-alone sites).

Multi-use developments are commonly found *ranging in size from 100,000 sq. ft. to 2 million sq. ft.* The data presented in this chapter correspond to multi-use developments in this size range. The recommended procedures for estimating trip generation at multi-use developments are likely applicable at even larger sites, but the analyst is encouraged to collect additional data.

A key characteristic of a multi-use development is that trips among the various land uses can be made on site and these *internal trips are* not *made on the major street system.* In some multi-use developments, these internal trips can be made either by walking or by vehicles using internal roadways without using external streets.

An *internal capture rate* can generally be defined as a percentage reduction that can be applied to the trip generation estimates for individual land uses to account for trips internal to the site. It is important to note that these reductions are applied externally to the site (i.e., at entrances, adjacent intersections

> **Multi-Use Development**
>
> ◆ Typically planned as a single real-estate project,
>
> ◆ Typically between 100,000 and 2 million sq. ft. in size,
>
> ◆ Contains two or more land uses,
>
> ◆ Some trips are between on-site land uses, and
>
> ◆ Trips between land uses do not travel on major street system.
>
> ***Not a(n)***
>
> ◆ Central business district,
>
> ◆ Suburban activity center, or
>
> ◆ Existing ITE land use classification with potential for a mix of land uses, such as
>
> • Shopping center,
>
> • Office park with retail,
>
> • Office building with retail, or
>
> • Hotel with limited retail and restaurant space.

and adjacent roadways). The trip reduction for internally captured trips is separate from the reduction for *pass-by trips.* These are two distinct phenomena and both could be applicable for a proposed development. The internal trips, if present, should be subtracted out *before* pass-by trip reductions are applied (refer to Chapter 5 for a complete discussion of pass-by trip estimation).

7.3 What Is *Not* a Multi-Use Development?

In literal terms, a multi-use development could mean any combination of different land use types within a defined, congruous area. But that definition would encompass a wide range of potential applications, many of which are not intended to be the focus of this chapter.

A traditional **downtown or central business district (CBD)** is <u>not</u> considered a multi-use development for purposes of this handbook. Downtown areas typically have a mixture of diverse employment, retail, residential, commercial, recreation and hotel uses. Extensive pedestrian interaction occurs because of the scale of the downtown area, ease of access and proximity of the various uses. Automobile occupancy, particularly during peak commuting hours, is usually higher in the CBD than in outlying areas. Some downtowns have excellent transit service. For these reasons, trip generation characteristics in a downtown environment are different from those found in outlying or suburban areas. The focus of the data presented throughout *Trip Generation* is on sites in suburban settings with limited or no transit service and free parking. ***Accordingly, trip generation characteristics in this chapter, and specifically in the case of capture rates at multi-use developments, are directly <u>applicable only to sites outside the traditional downtown.</u>*** The potential effects of transit service and on-site parking fees are discussed in Appendix B.

A **shopping center** could also be considered a multi-use development. However, because data have been collected directly for them, shopping centers are considered in *Trip Generation* as a single land use. The associated trip generation rates and equations given in *Trip Generation* reflect the "multi-use" nature of the development because of the way shopping center data have been collected. ***Accordingly, internal capture rates are not applicable and <u>should not be used to forecast trips for shopping centers</u> if using statistis and data for Land Use Code 820.*** However, if the shopping center is planned to have out-parcel development of a significantly different land use classification or a very large percentage of overall GLA, the site could be considered a multi-use development for the purpose of estimating site trip generation.

Likewise, a subdivision or planned unit development containing general office buildings and support services such as banks, restaurants and service stations arranged in a park- or campus-like atmosphere should be considered as an **office park** (Land Use Code 750), not as a multi-use development. Similarly, office buildings with support retail or restaurant facilities contained inside the building should be treated as **general office buildings** (Land Use Code 710) because the trip generation rates and equations already reflect such support uses. A hotel with an on-site restaurant and small retail falls within Land Use Code 310 and should not be treated as a multi-use development.

7.4 Methodology for Estimating Trip Generation at Multi-Use Sites

Internally captured trips can be a significant component in the travel patterns at multi-use developments. However, more studies are needed to thoroughly quantify this phenomenon. Section 7.5 presents a recommended procedure for estimating internal capture rates (and a worksheet for organizing and documenting the analysis assumptions used in the estimation of the internal capture rates) for multi-use development sites.

The internal trip-making characteristics of multi-use development sites are directly related to the mix of on-site land uses (which are typically a combination of residential, office, shopping/retail, restaurant, entertainment and hotel/motel). When combined within a single mixed-use development, these land uses tend to interact and thus attract a portion of each other's trip generation.

The recommended methodology for estimating internal capture rates and trip generation at multi-use sites is based on two fundamental assumptions. First, the proportions of trips between interacting land use types (which will be satisfied internally by pairs of land uses) are assumed to be relatively stable. Second, if sufficient data were available, these internal capture percentages could be predicted with adequate confidence. The need for additional data collection at multi-use developments is described in Section 7.7.

As should be expected, the observed internal capture rates for multi-use developments vary by time of day, the site's mix of land uses and size of the development.

Several premises frame the recommended methodology. An example to illustrate its application is presented in the highlighted text to the side. Key to the success of this methodology in replicating internal capture patterns at multi-use sites is its iterative, balancing steps that constrain internal trip-making levels to what are realistic given the mix of land uses.

Illustration of Methodology Overview

Assume a multi-use development with a mix of office, retail and residential uses. Assume that the office building generates 500 exiting trips during the evening peak hour (based on factors presented in *Trip Generation*).

Based on surveys at other multi-use developments (for illustration purposes), it is estimated that the 500 peak hour trips could go to the following destinations: 5 trips to another office building within the development, 115 trips to a retail site within the development, 10 trips to residential units on-site and 370 to external sites (as illustrated in Figure 7.1a).

What if there are no on-site residential units? The number of trips from the office to an internal residential destination changes to zero and the number of trips to external destinations becomes 380 (i.e., the total trips from the office building remains constant at 500).

What if there are a large number of on-site residences? Assume the residential uses generate 1,000 entering trips during the evening peak hour. As illustrated in Figure 7.1b, the trips are assumed to originate as follows: 20 trips from an on-site office building, 310 trips from on-site retail, no trips from another on-site residential component and 670 trips from external origins.

With the larger number of residences, as many as 20 trips could come from on-site office buildings. But the actual on-site office buildings generate only 10 trips to the on-site residential land use. So, 10 trips would be expected from on-site office to on-site residential in Figure 7.1c. The key assumption is that the *"balanced" number of internal trips will match the controlling (i.e., lower) value.*

Figure 7.1 Illustration of Internal Trip Balancing for a Multi-Use Development

DISTRIBUTION OF POTENTIAL DESTINATIONS OF TRIPS **FROM OFFICE** USE

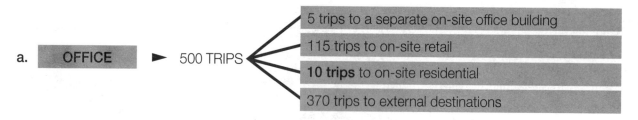

a. OFFICE ► 500 TRIPS

5 trips to a separate on-site office building

115 trips to on-site retail

10 trips to on-site residential

370 trips to external destinations

b. DISTRIBUTION OF POTENTIAL ORIGINS OF TRIPS **TO RESIDENTIAL** USE

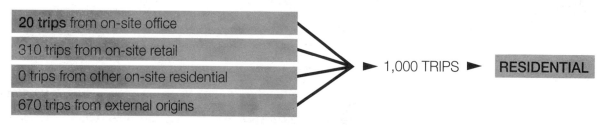

20 trips from on-site office

310 trips from on-site retail

0 trips from other on-site residential

670 trips from external origins

► 1,000 TRIPS ► RESIDENTIAL

c. BALANCED[1] DISTRIBUTION OF ORIGINS OF TRIPS **TO RESIDENTIAL** USE

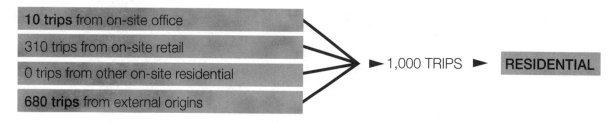

10 trips from on-site office

310 trips from on-site retail

0 trips from other on-site residential

680 trips from external origins

► 1,000 TRIPS ► RESIDENTIAL

[1] Only the office-to-residential values have been "balanced." Similarly, all other land use pairs would need to be balanced.

Premise 1: The distribution of trip purposes among motorists entering or exiting a development site is relatively stable. The distribution of destination land uses is likewise assumed to be relatively stable. For example, the destinations of trips from an office building are distributed among the many potential destinations (e.g., retail, residential, other office) in roughly the same pattern whether the office is stand-alone or in a multi-use development.

Premise 2: The converse of Premise 1 is also true, that the distribution of origins for trips to a particular land use is relatively stable.

Premise 3: The number of trips *from* a land use within a multi-use development *to* another land use within the same multi-use development (i.e., an internal trip) is a function of the size of the "receiving" land use and the number of trips it attracts, as well as the size of the "originating" land use and the number of trips it sends. *The number of trips between a particular pair of internal land uses is limited to the smaller of these two values.*

7.5 Procedure for Estimating Multi-Use Trip Generation

The recommended procedure for trip estimation, although complex, simplifies the actual trip-making dynamics within a multi-use development. For example, the procedure does not take into account a number of key variables that are likely to affect the internal capture rate, such as proximity of on-site land uses (and pedestrian connections between them) and location of the multi-use site within the urban/suburban area (and the proximity of competing or complementary land uses). **The analyst is encouraged to exercise caution in applying the data presented herein because of the limited sample size and scope. Additional data should be collected where possible (refer to Section 7.7 for guidance). The analyst is also encouraged to make logical assumptions in his/her use of this procedure. In summary, use good professional judgment.**

WORDS OF ADVICE

◆ Collect additional data if possible

◆ Exercise caution

◆ Be logical

◆ Use good professional judgment

The step-by-step procedure, as described below, can be used for any number of land uses within the multi-use site. Sample forms are provided for three and four land uses, however, the analyst can modify the sample worksheet to correspond to the desired number of land uses. The layout of the worksheet will become more complex as additional land uses are included.

Blank worksheets for estimating multi-use development trip generation are provided at the end of this chapter. The following step-by-step procedures illustrate how the worksheet should be completed.

Step 1. Document Characteristics of Multi-Use Development

Enter the following information onto the worksheet:

◆ Name of development;

◆ Description of each land use in the development and its ITE land use code; and

◆ Size of each land use, corresponding to the most appropriate independent variable used in *Trip Generation* (e.g., gross leasable area, gross floor area, dwelling units).

If the site has two or more buildings containing the same land use, combine the sizes of the multiple buildings if they are situated within reasonable and convenient walking distance of each other. If the buildings are not close to each other, treat them as separate land uses on the worksheet (for example, as Office A and Office B).

If the site has multiple residential components (single-family, apartment, etc.), compute the trip generation for each residential type separately (later in Step 3), but record as only a single land use on the worksheet.

Step 2. Select Time Period for Analysis

Enter the time period for which the analysis is being conducted onto the worksheet (for help in selecting the appropriate time period for analysis, refer to Chapter 2 of this handbook).

Internal capture rates vary by time of day. A separate worksheet should be completed for each distinct time period. It should be noted that typical internal capture rates are presented later in this chapter for the weekday midday, weekday evening peak and weekday daily.

Internal capture rates may also vary by day of the week. The typical internal capture rates used in a later step are based on data collected on a Tuesday, Wednesday, or Thursday (unless specifically noted otherwise). Analyses for a Friday or Saturday may need modified rates.

Step 3. Compute Baseline Trip Generation for Individual Land Uses

Compute the number of trips generated for the desired time period for each land use based on the given independent variable.

◆ Refer to notes in Step 1 if there are multiple buildings of the same land use within the site.

◆ Compute number of trips generated by direction (enter/exit).

◆ Use the *Trip Generation* rate, *Trip Generation* equation, or local data for each land use. Refer to Chapter 3 for guidance on how to select the appropriate rate or equation for each land use. Do not adjust for pass-by or diverted linked trips at this time.

Record trip generation values in worksheet. For each land use, record the baseline trip generation in the column under the "total" heading.

SAMPLE PROBLEM

Step 1. For our example problem, we are analyzing a multi-use site comprised of a 200,000-sq. ft. shopping center; a 120,000-sq. ft. office building; and 200 low-rise apartments. On the worksheet in Figure 7.2, the three land use types and their corresponding ITE land use codes and sizes are recorded.

Step 2. We will assume the analysis time period is the evening peak hour of adjacent street traffic (as indicated in the worksheet in Figure 7.2).

Step 3. For Land Use Code 820, use the equation from page 1,453 of *Trip Generation*, Seventh Edition, to compute trips; for Land Use Code 710, use the equation from page 1,160; for Land Use Code 221, use the equation from page 337. The results are listed in the worksheet in Figure 7.2.

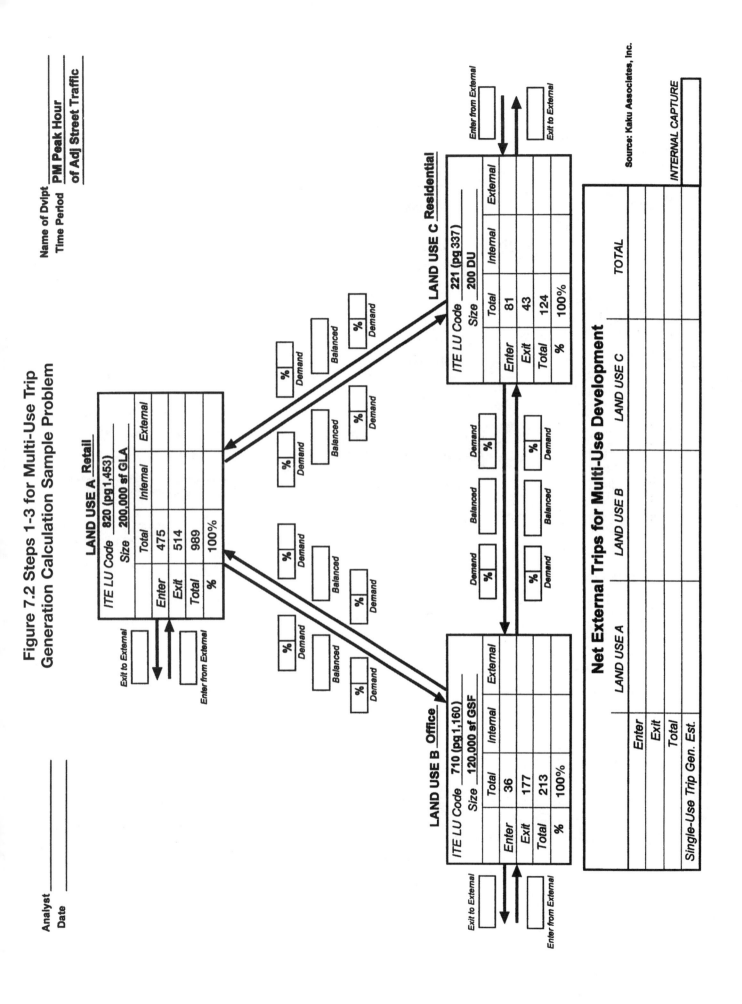

Figure 7.2 Steps 1-3 for Multi-Use Trip Generation Calculation Sample Problem

Step 4. Estimate Anticipated Internal Capture Rate Between Each Pair of Land Uses

Tables 7.1 and 7.2 present unconstrained internal capture rates that have been estimated on the basis of a series of studies conducted in Florida. These are the only data available to ITE prior to publication that are detailed enough for credible use. Readers are encouraged to collect and submit additional data to ITE using procedures described in Section 7.7. As the best available applicable data, it is recommended that these internal capture rates be used unless local data are collected.

SAMPLE PROBLEM (continued)

Step 4. The sample worksheet in Figure 7.3 shows the recorded "internal capture" rates for each pair of land uses.

Estimate the interaction between each pair of land uses for the selected time period.

◆ Use Tables 7.1 and 7.2 (or local data) as the basis for the estimate. (Note: there are no data provided for the weekday morning peak period or for the Saturday midday peak period.)

◆ Table 7.1 presents estimated unconstrained internal capture

rates for trip origins within a multi-use development. For example, during the weekday midday peak period, of all the vehicle-trips **exiting** an on-site office use, 2 percent of the trips could be destined for another on-site office use and 20 percent destined for on-site retail use.

◆ Table 7.2 presents estimated unconstrained internal capture rates for trip destinations within a multi-use development. For example, during the weekday midday peak period, of all the vehicle-trips entering an on-site retail use, 4 percent of the trips could originate at an on-site office use and 5 percent at an on-site residential use.

Record the estimated capture rates on the worksheet (in the boxes marked "demand").

◆ For each land use pairing, record four values; for example, for the pairing of retail and office uses, the following four values should be recorded:
  • Percent of trips from on-site office destined to an internal retail destination
  • Percent of trips to on-site retail originating from an internal office use
  • Percent of trips from on-site retail destined to an internal office destination
  • Percent of trips to on-site office originating from an internal retail use

◆ Each value represents the unconstrained demand (or maximum potential trip interaction between the two land uses), by direction.

Because of the limited database on trip characteristics at multi-use sites, the analyst is cautioned to review the particular characteristics of the multi-use development under analysis before using the factors presented in Tables 7.1 and 7.2. Specifically, the analyst must *assess whether each set of internal trip capture rates makes sense considering the particular individual land uses within the multi-use development.*

If local data on internal capture rates by land use pair can be obtained, the local data should be given preference.

The data in Table 7.1 are limited to trip interaction among the three land uses for which sufficient data were available. *If an on-site land use does not match a land use category in Table 7.1, either (1) collect local data to establish an internal capture rate, according to procedures described in Section 7.7 of this chapter, or (2) assume no internal capture.* (Note: although this assumption of no internal capture may be unrealistic, in the absence of any data it is better to overestimate off-site vehicle-trips.)

Table 7.1 Unconstrained Internal Capture Rates for Trip Origins within a Multi-Use Development

		WEEKDAY		
		MIDDAY PEAK HOUR	p.m. PEAK HOUR OF ADJACENT STREET TRAFFIC	DAILY
from OFFICE	to Office	2%	1%	2%
	to Retail	20%	23%	22%
	to Residential	0%	2%	2%
from RETAIL	to Office	3%	3%	3%
	to Retail	29%	20%	30%
	to Residential	7%	12%	11%
from RESIDENTIAL	to Office	N/A	N/A	N/A
	to Retail	34%	53%	38%
	to Residential	N/A	N/A	N/A

Caution: The estimated typical internal capture rates presented in this table rely directly on data collected at a limited number of multi-use sites in Florida. While ITE recognizes the limitations of these data, they represent the only known credible data on multi-use internal capture rates and are provided as illustrative of typical rates. ***If local data on internal capture rates by paired land uses can be obtained, the local data may be given preference.***

N/A—Not Available; logic indicates there is some interaction between these two land uses; however, the limited data sample on which this table is based did not record any interaction.

Table 7.2 Unconstrained Internal Capture Rates for Trip Destinations Within a Multi-Use Development

| | | WEEKDAY | | |
		MIDDAY PEAK HOUR	p.m. PEAK HOUR OF ADJACENT STREET TRAFFIC	DAILY
to OFFICE	from Office	6%	6%	2%
	from Retail	38%	31%	15%
	from Residential	0%	0%	N/A
to RETAIL	from Office	4%	2%	4%
	from Retail	31%	20%	28%
	from Residential	5%	9%	9%
to RESIDENTIAL	from Office	0%	2%	3%
	from Retail	37%	31%	33%
	from Residential	N/A	N/A	N/A

Caution: The estimated typical internal capture rates presented in this table rely directly on data collected at a limited number of multi-use sites in Florida. While ITE recognizes the limitations of these data, they represent the only known credible data on multi-use internal capture rates and are provided as illustrative of typical rates. **If local data on internal capture rates by paired land uses can be obtained, the local data may be given preference.**

N/A—Not Available; logic indicates there is some interaction between these two land uses; however, the limited data sample on which this table is based did not record any interaction.

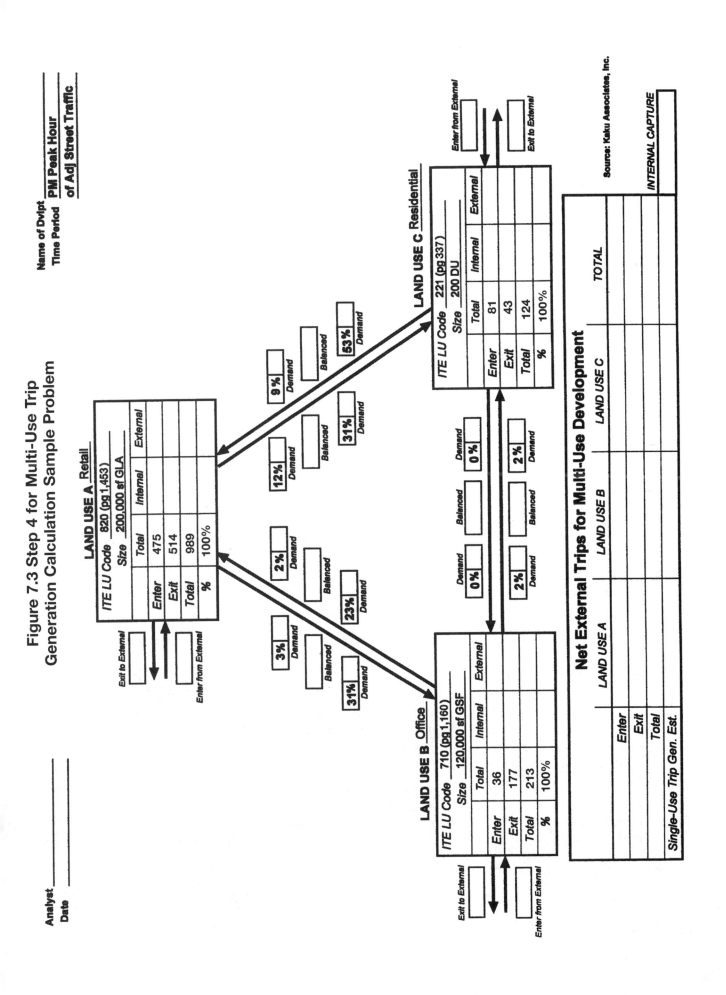

Figure 7.3 Step 4 for Multi-Use Trip Generation Calculation Sample Problem

Step 5. Estimate "Unconstrained Demand" Volume by Direction

Multiply the internal capture percentages by the appropriate directional trip generation value in the worksheet.

◆ For each pair of land uses, compute a directional value from the percentages that were entered. (Note: these values will be balanced later in Step 6.)

Record the "unconstrained demand" volumes by direction on the worksheet in the boxes marked "demand" next to the percentages.

Step 6. Estimate "Balanced Demand" Volume by Direction

Compare the two values in each direction for each land use pairing and **select the lower (i.e., controlling) value.**

Record the value as the "balanced demand" (the lower of the directional internal volumes) between each pair of land uses.

◆ Record the lower value for each land use for each direction

◆ Record in the worksheet boxes marked "balanced."

Step 7. Estimate Total Internal Trips to/from Multi-Use Development Land Uses

For each land use, first sum the internal trips *to* each other land use. Then for each land use, sum the internal trips *from* each other land use. Record these total internal trip values in the worksheet in the summary table for each land use.

Compute and record the internal percentages for each land use in the summary table for each land use. **Review values and verify that they are reasonable.**

SAMPLE PROBLEM
(continued)
Step 5. The "unconstrained demand" volumes are computed by multiplying the directional trip generation value by the "unconstrained demand" percentage, as shown in the sample worksheet in Figure 7.4. For example,

◆ Trips from retail to office: 514 outbound trips × 3% = 15 trips

◆ Trips to office from retail: 36 inbound trips × 31% = 11 trips

◆ Trips from office to retail: 177 outbound trips × 23% = 41 trips

◆ Trips to retail from office 475 inbound trips × 2% = 10 trips

---

Step 6. Select the controlling value (i.e., the lower value) for each pair of land uses for each direction. For example, in the Figure 7.4 worksheet,

◆ For trips from retail to office, the retail could generate 15 internal trips but the office could only receive 11 internal trips; the controlling value is 11 internal trips.

◆ For trips from office to retail, the office could generate 41 internal trips but the retail could only receive 10 internal trips; the controlling value is 10 internal trips.

Step 7. The sample worksheet in Figure 7.5 illustrates Step 7. For the retail land use, 10 internal trips are estimated from the on-site office and 23 internal trips from the on-site residential. Therefore, the total internal trips entering the retail land use is 33. The internal trips exiting retail sum to 36 (11 to the on-site office and 25 to the on-site residential). In total, seven percent of the retail trips (69 of 989) are internal to the multi-use site. This procedure is followed for each land use.

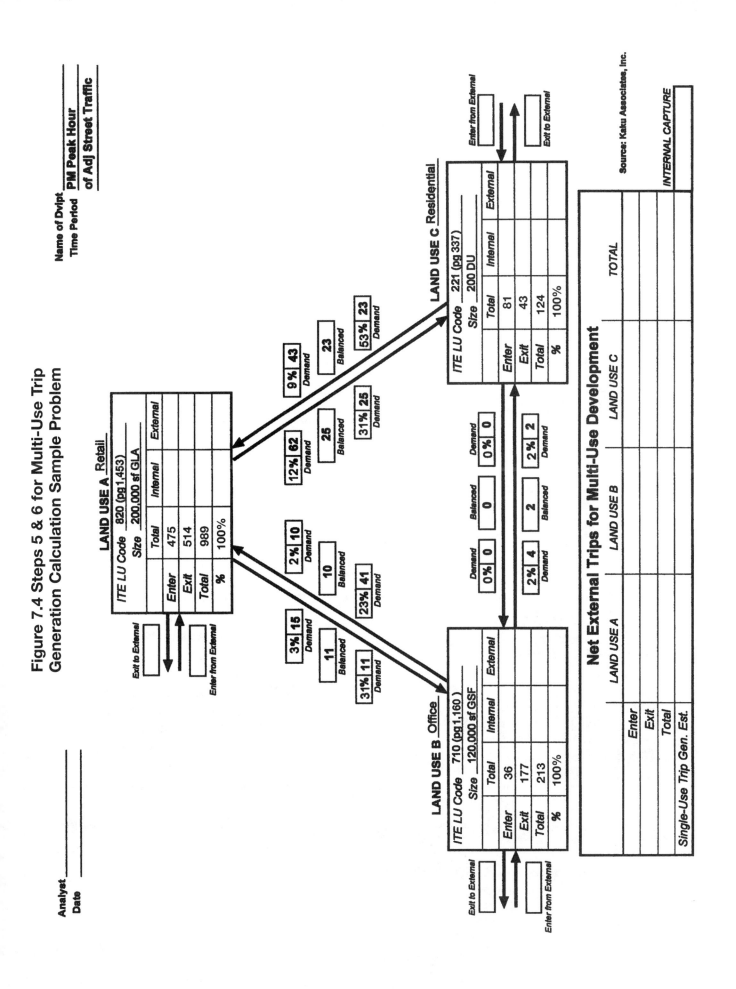

Figure 7.4 Steps 5 & 6 for Multi-Use Trip Generation Calculation Sample Problem

Step 8. Estimate the Total External Trips for Each Land Use

Calculate the number of external trips (by direction) by subtracting the estimated internal trips from the total trips for each individual land use. Record values in tables for each land use and in the boxes marked "exit to external" and "enter from external."

Step 9. Calculate Internal Capture Rate and Total External Trip Generation for Multi-Use Site

Record the final external trip estimates for each land use onto the worksheet and in the table of "net external trips."

Compute the *net external trip generation* for the entire site by summing the external volumes for each of the site land uses.

Record the original estimates for total trip generation for each land use onto the worksheet in the row denoted "original trip generation estimate." Compute the overall *internal capture rate* by dividing the net external trip generation estimate by the original total trip generation estimate, and subtracting the quotient from 100 percent.

7.6 Cautions Regarding Recommended Procedure

The data presented in Section 7.5 quantify the influence of several key factors on internal capture rates. Numerous other factors have a direct influence on travel at multi-use sites, factors for which the current data do not account. Additional data and analysis are desirable to better quantify the relationships between these factors and multi-use development trip generation and internal capture rates. A summary description of the pertinent information contained in several existing documents is included in Appendix C of this handbook.

Limited Sample Size—The estimated typical internal capture rates presented in Section 7.5 in Tables 7.1 and 7.2 rely directly on data collected at a limited number of multi-use sites in Florida. While ITE recognizes the limitations of these data, they represent the only known credible data on multi-use internal capture rates and are provided as illustrative of typical rates. *If local data on internal capture rates by land use pair can be obtained, the local data should be used (and the data submitted to ITE for use in future publications).*

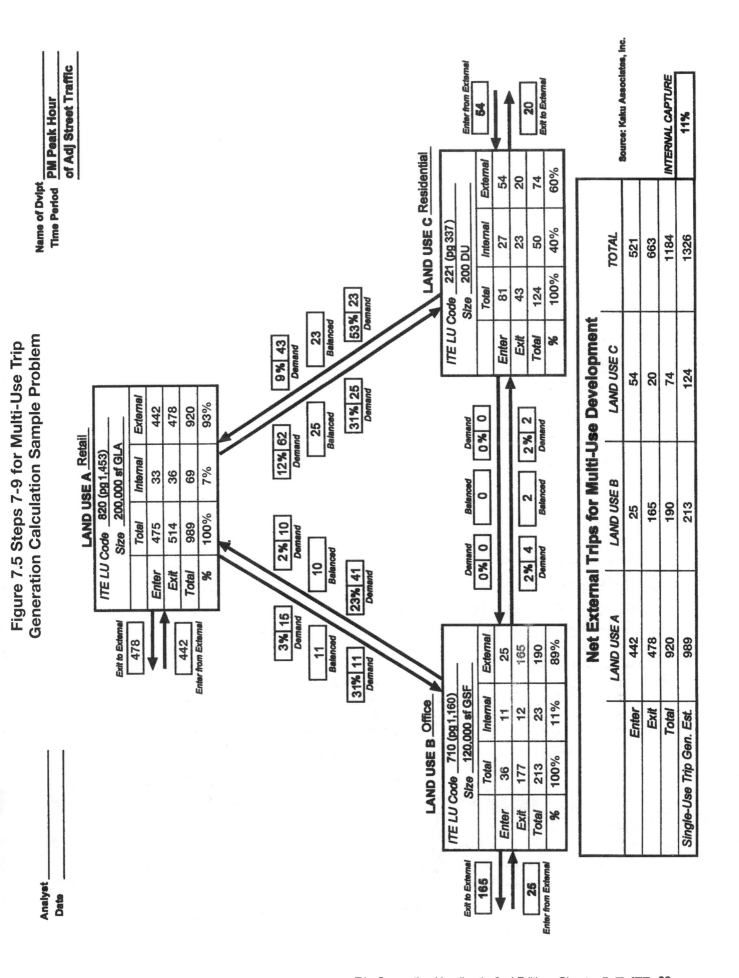

Figure 7.5 Steps 7-9 for Multi-Use Trip Generation Calculation Sample Problem

LAND USE A Retail

ITE LU Code 820 (pg 1,453) Size 200,000 sf GLA

	Total	Internal	External
Enter	475	33	442
Exit	514	36	478
Total	989	69	920
%	100%	7%	93%

Exit to External **478**
Enter from External **442**

LAND USE C Residential

ITE LU Code 221 (pg 337) Size 200 DU

	Total	Internal	External
Enter	81	27	54
Exit	43	23	20
Total	124	50	74
%	100%	40%	60%

Enter from External **54**
Exit to External **20**

9% **43** Demand
23 Balanced
53% **23** Demand

12% **62** Demand
25 Balanced
31% **25** Demand

Demand **0%** **0**
Balanced **0**
Demand **0%** **0**

Demand **2%** **4**
Balanced **2**
Demand **2%** **2**

LAND USE B Office

ITE LU Code 710 (pg 1,160) Size 120,000 sf GSF

	Total	Internal	External
Enter	36	11	25
Exit	177	12	165
Total	213	23	190
%	100%	11%	89%

Exit to External **165**
Enter from External **25**

2% **10** Demand
10 Balanced
23% **41** Demand

3% **15** Demand
11 Balanced
31% **11** Demand

Net External Trips for Multi-Use Development

	LAND USE A	LAND USE B	LAND USE C	TOTAL
Enter	442	25	54	521
Exit	478	165	20	663
Total	920	190	74	1184
Single-Use Trip Gen. Est.	989	213	124	1326

Source: Kaku Associates, Inc.

INTERNAL CAPTURE **11%**

Pass-By Trips—The application of pass-by trip reductions presented in Chapter 5 should be likewise applicable to multi-use sites. However, none of the internal trips can be of a pass-by nature because they do not travel on the adjacent (external) street system. *Pass-by trip percentages are applicable only to trips that enter or exit the adjacent street system.* Use the pass-by trip estimation procedure in Chapter 5 of this handbook.

Competing Markets—Proximity to competing markets is expected to influence internal capture rates. *The greater the distance to external competing uses, the greater the likelihood of capturing trips internally within a multi-use development site.* Developments in a suburban community may have higher capture rates than those in urban developments since urban areas provide a higher number of alternative opportunities than many suburban developments. For example, residents in an urban mixed-use development have more choices in shopping opportunities and thus may travel outside the development site for their shopping needs, even though there are retail uses in their development site. Suburban residents, on the other hand, may not have as many alternative opportunities and therefore may be more likely to confine their trips to the mixed-use site for their shopping or other needs. However, at this time there are no site-trip generation data available on which to base adjustment factors of this type.

Proximity and Density of On-Site Land Uses—The proximity and density of the residential, retail, office and hotel uses will affect internal trip-making. *Generally, the greater the density and the closer the proximity of the complementary uses on site, the greater the level of internalization of trips.* The proximity should be measured in terms of both distance and impedance to the traveler. For example, the presence of foot paths or bicycle paths, protected crosswalks or overpasses and pedestrian refuge areas greatly enhance the accessibility of paired on-site land uses. At this time, however, no site-trip generation data are available on which to base adjustment factors of this type.

Other Site-Specific Issues—Many other issues potentially affect trip making at multi-use sites. For example, can those who work on-site afford to live on-site? How long will it take for the office uses to attract work trips from on-site residences? Is there an internal circulation system that enhances or discourages internal trips?

Shared Parking—Shared parking and multi-use trip generation estimation methodologies, though similar, are not interchangeable. Shared parking factors cannot be applied to estimate trip generation at multi-use developments.

7.7 Data Collection at Multi-Use Developments

ITE wishes to increase the database on multi-use developments in order to provide internal capture data for a broader range of land uses. ITE would appreciate additional data from analyses of such developments.

A data collection program for a multi-use development site should include verification that the site to be surveyed is appropriate for inclusion in the ITE database. It should also include a compilation of information describing site characteristics and field data collection. The field data collection at the site should have at least two components: in-person interviews on-site and a cordon traffic count. Conducting internal traffic counts should be considered at sites where internal streets exist and can be isolated and where internal streets carry most of the internal trips (both pedestrian and vehicular).

A data collection program that has all three components will provide the clearest understanding of internal and external trip-making at the multi-use site. If only an on-site interview is conducted, factoring of the survey results to calibrate to actual trip volumes will not be possible. In general, experience has shown that data collection efforts consisting solely of interviews tend to overstate the actual proportion of internal trips at a multi-use site.

At a minimum, data collection should consist of on-site, in-person interviews coupled with a complete cordon count.

Site Selection

The site should be fully developed, operational and mature. New or partially developed sites may not generate trips (both external and internal) at the rate expected of a mature site. (Note: the degree of occupancy is one of the site characteristics to be collected, as described below.)

The driveways serving the multi-use site should not serve any adjacent property. If driveways are shared with another site, it is not possible to count the traffic destined for the multi-use site using traditional traffic counting methods. In addition, the selected multi-use site should have a minimal presence of through trips (i.e., external trips that pass through the site without stopping).

Site Characteristics

Compile the necessary information to describe the multi-use development and each of its individual land uses. At a minimum, obtain information on the independent variables reported under each of the individual land uses in *Trip Generation*. For example, this would include, as appropriate, gross square footage (total and occupied), employees, hotel rooms (total and occupied), dwelling units (total and occupied), restaurant seats, presence of drive-through windows and

so forth. A map or sketch should also be prepared showing buildings, internal streets, access to the external street system and locations of counts and interviews.

If possible, the data collection program should obtain a description and assessment of the proximity/accessibility of complementary uses within the site and a description and assessment of the proximity of competing markets outside the site.

Traffic Counts (Cordon)

Driveway volumes at all entrances/exits at the multi-use site should be counted for as long of a period as possible. If only 48 hours of data can be obtained, volumes should be counted during the mid-week (Tuesday through Thursday) to avoid daily variations that may occur on Monday and Friday. If the selected period for design of site access could be the weekend, traffic counts and surveys should likewise be conducted during the weekend.

Ideally, 7 consecutive days of data are recommended if budgets allow and if site driveways are configured to enable complete and accurate counts. With 7 days of data, daily variations can be computed and a weekday average and weekend average can be calculated. Driveway counts should be conducted during the same periods as interviews.

Traffic Counts (Internal)

For some multi-use developments, it will be possible to validate the survey results for overall internal trip-making with a comprehensive count of internal pedestrians and vehicles. In such cases, pedestrians and vehicles traveling among on-site land uses should be counted during the interview time periods.

Interviews

Concurrent with gathering driveway volumes, interviews of workers, shoppers, visitors and residents of the site should be conducted. In general, the objective of the intercept survey is to obtain information on trip purposes at the multi-use site, the origins and destinations of trips entering and exiting the site and the mode of each trip.

Interviews of persons are typically conducted on site as they leave the site (or leave a single land use within the site). Each interview obtains information on the trips to and from the site. A sample list of interview questions is provided in Figures 7.6 and 7.7. The questions are written for on-site administration of the survey. If the survey will be conducted at the cordon driveway, the analyst will need to revise the questions to account for potential multiple on-site stops.

The actual field survey form should also include a space for the interviewer to record the date, name of the development, interviewer's location within the site,

time of the interviews (half-hour intervals should be sufficient) and interviewer's name.

A minimum of 100 interviews per time period should be conducted at the multi-use development. For larger developments (i.e., with at least 300,000 sq. ft. of office or retail space), a minimum of 200 interviews per time period should be completed.

Submittal of Multi-Use Development Data to ITE

A summary of the survey and traffic count results should be submitted to ITE for use in updating the multi-use development database and trip estimation methodology in subsequent updates to the *Trip Generation Handbook*. The report should include a **description of the site and its setting**, a summary of the **data collection program**, the measured internal capture rates between on-site land uses, and a comparison of the **actual external trip generation of the site** to the sum of the trip generation estimates for individual uses within the site. It is strongly suggested that all trip generation studies for multi-use developments follow the procedures presented in this chapter.

Tables 7.3 and 7.4 present a suggested tabulation of multi-use internal capture data. Values should represent factored numbers summed from all of the interview stations. Table 7.3 summarizes the distribution of trip origins and destinations

for *trips heading to land uses within the multi-use development*.

If the survey instrument and format shown in Figure 7.6 is used, the "from" end of the trip is compiled from answers to Question C; the "to" end of the trip is compiled from responses to Question B. The analyst should also compile information on trips "from" the land use described in Question B to an on-site land use identified in Question A.

If the survey instrument and format shown in Figure 7.7 is used, the "from" end of the trip is compiled from answers to Question C; the "to" end of the trip is compiled from responses to Question A that indicate an on-site destination.

Table 7.4 summarizes the distribution of trip origins and destinations *for trips heading from land uses within the multi-use development*.

If the survey instrument and format shown in Figure 7.6 is used, the "from" end of the trip is compiled from answers to Question B; the "to" end of the trip is compiled from responses to Question A.

If the survey instrument and format shown in Figure 7.7 is used, the "from" end of the trip is compiled from answers to Question A; the "to" end of the trip is compiled from responses to Question B.

The pass-by data (Question D) should be summarized as a single value across all trip purposes.

Send the multi-use development study results to:

Institute of Transportation Engineers
1099 14th Street, NW
Suite 300 West
Washington, DC 20005-3438
Tel: +1 202-289-0222
Fax: +1 202-289-7722

Figure 7.6 Suggested Interview Questions, Multi-Use Development

AS PERSONS DEPART (Preferred Survey)

A. WHERE ARE YOU HEADED? (TWO-PART QUESTION; CIRCLE ONE IN EACH)

1. Within this multi-use site   2. Outside this site

a. Office	e. Medical/Office	i. Other (Specify):_____
b. Retail	f. Hotel/Motel	_____
c. Restaurant	g. Residential	
d. Bank	h. Theater/Entertainment	

B. WHERE ARE YOU COMING FROM (I.E., ON SITE)? (CIRCLE ONLY ONE)

a. Office	e. Medical/Office	i. Other (Specify):_____
b. Retail	f. Hotel/Motel	_____
c. Restaurant	g. Residential	
d. Bank	h. Theater/Entertainment	

C. BEFORE THAT, WHERE DID YOU COME FROM? (TWO-PART QUESTION; CIRCLE ONE IN EACH)

1. Within this multi-use site   2. Off site

a. Office	e. Medical/Office	i. Other (Specify):_____
b. Retail	f. Hotel/Motel	_____
c. Restaurant	g. Residential	
d. Bank	h. Theater/Entertainment	

D. WOULD YOU HAVE DRIVEN BY THIS SITE IF YOU DID NOT STOP HERE THIS [AFTERNOON/MORNING]?

1. Yes   2. No

Figure 7.7 Suggested Interview Questions, Multi-Use Development

AS PERSONS ARRIVE (Use only if exit surveys are not possible)

A. WHERE ARE YOU HEADED? (CIRCLE ONLY ONE)

a. Office	e. Medical/Office	i. Other (Specify):_____
b. Retail	f. Hotel/Motel	_____
c. Restaurant	g. Residential	
d. Bank	h. Theater/Entertainment	

B. AFTER THAT, WHERE WILL YOU GO? (TWO-PART QUESTION; CIRCLE ONE IN EACH)

1. Within this multi-use site   2. Off site

a. Office	e. Medical/Office	i. Other (Specify):_____
b. Retail	f. Hotel/Motel	_____
c. Restaurant	g. Residential	
d. Bank	h. Theater/Entertainment	

C. WHERE DID YOU COME FROM? (TWO-PART QUESTION; CIRCLE ONE IN EACH)

1. Within this multi-use site   2. Off site

a. Office	e. Medical/Office	i. Other (Specify):_____
b. Retail	f. Hotel/Motel	_____
c. Restaurant	g. Residential	
d. Bank	h. Theater/Entertainment	

D. WOULD YOU HAVE DRIVEN BY THIS SITE IF YOU DID NOT STOP HERE THIS [AFTERNOON/MORNING]?

1. Yes   2. No

Table 7.3 Suggested Tabulation of Multi-Use Internal Capture Data Trips <u>TO</u> On-Site Land Uses

FROM	TO OFFICE	TO RETAIL	TO RESTAURANT	TO BANK	TO MEDICAL	TO HOTEL	TO RESIDENTIAL	TO THEATER	TO _____	TOTAL
on-site										
off-site										
on-site										
office										
retail										
restaurant										
bank										
medical										
hotel										
residential										
theater										
other										
off-site										
office										
retail										
restaurant										
bank										
medical										
hotel										
residential										
theater										
other										
Total										

Table 7.4 Suggested Tabulation of Multi-Use Internal Capture Data Trips <u>FROM</u> On-Site Land Uses

TO	FROM OFFICE	FROM RETAIL	FROM RESTAURANT	FROM BANK	FROM MEDICAL	FROM HOTEL	FROM RESIDENTIAL	FROM THEATER	FROM _____	TOTAL
on-site										
off-site										
on-site										
office										
retail										
restaurant										
bank										
medical										
hotel										
residential										
theater										
other										
off-site										
office										
retail										
restaurant										
bank										
medical										
hotel										
residential										
theater										
other										
Total										

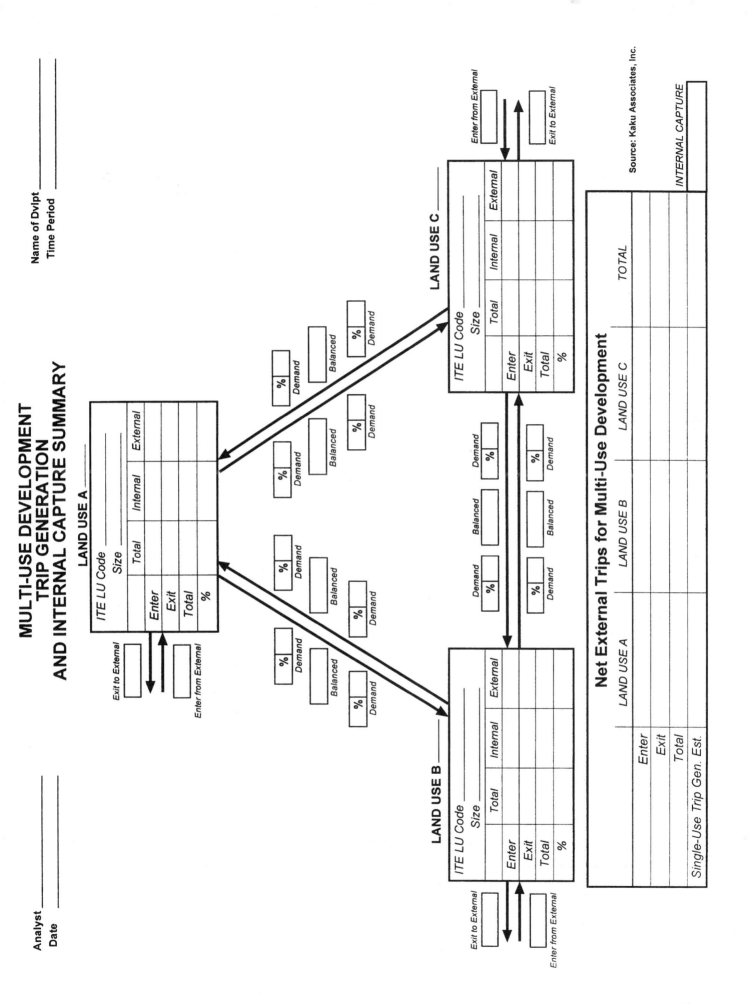

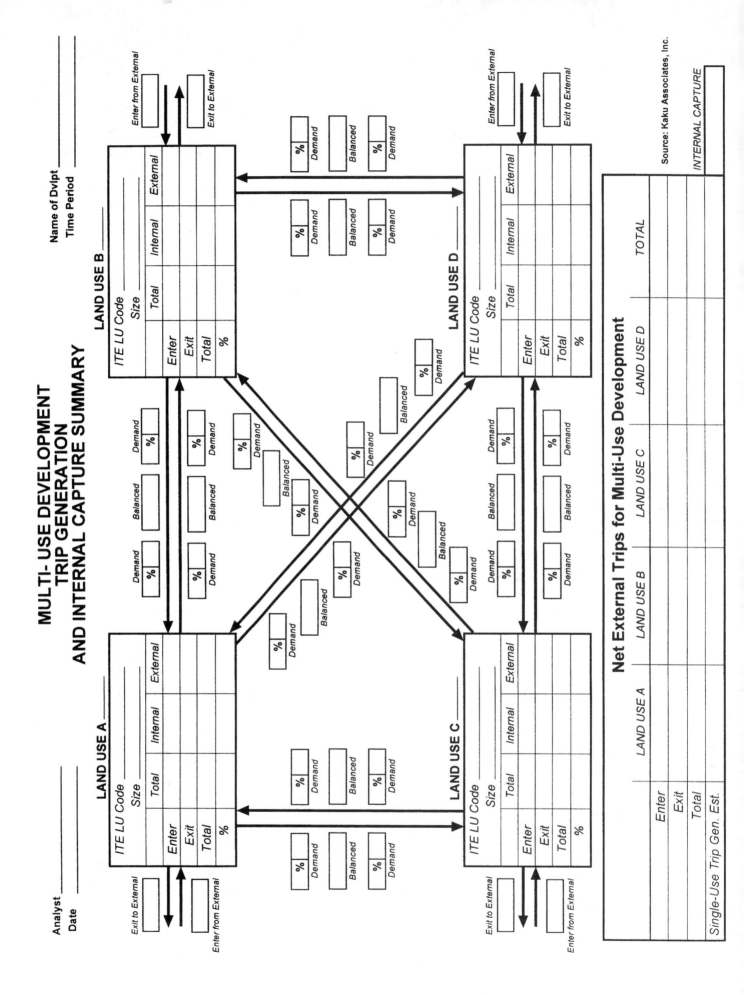

MULTI-USE DEVELOPMENT TRIP GENERATION AND INTERNAL CAPTURE SUMMARY

Truck Trip Generation

This appendix includes material that is strictly for informational purposes. It provides no recommended practices, procedures, or guidelines.

A.1 Background

The data in *Trip Generation* represent all vehicles entering or leaving a particular site. The analyst may, on occasion, need to explicitly estimate truck trips generated by a development site on a daily or peak hour basis. This appendix presents a summary of relevant literature that provides truck trip generation data.

Land Uses Expected to Generate Regular and Significant Volumes of Truck Traffic

◆ Landfills

◆ Truck terminal locations

◆ Truck stops and large gas/fuel stations

◆ Construction-related uses
  • Batch plants
  • Asphalt plants
  • Gravel pits

◆ Post office and parcel delivery companies

◆ State/county road commission offices/garages

◆ Any commercial or industrial uses where the aggregation of activity produces a significant goods-delivery traffic flow (shopping centers)

The potential needs for reasonably accurate estimates of truck trips fall into three general categories: traffic operations, street and road design, and public and political concerns.

Traffic operations are directly affected by the presence of trucks in the traffic stream. For example, levels of service for both roadway sections and intersections are affected by the percentages of heavy vehicles in the traffic stream. These effects can be exacerbated when other factors are introduced (e.g., if topography is severe, heavy vehicles will have an even greater negative effect).

Design considerations that need to be addressed with the aid of truck traffic data include both pavement design and the geometric design of the street or roadway. For example, if a new development generates a significant volume of trucks, the design of the roadway may be affected or, alternatively, traffic may need to be routed to different roadways in order to minimize pavement deterioration or adverse land use impacts. In addition to making sure that the roadway can handle increased loads from trucks, it is possible that estimates of truck traffic (in addition to a straightforward consideration of traffic volumes) may be needed to develop cost-sharing arrangements for the infrastructure improvements required for new development.

The analyst may need to address the geometric design of intersections, driveways and parking and service areas. If trucks are to be a regular and significant part of the traffic generated by a development, then the differential performance characteristics of heavy vehicles (e.g., turning radii and rates of braking and acceleration) must be considered.

Public and political concerns about the traffic impacts of development are often debated in public hearings and meetings (e.g., zoning commission meetings) and the popular press. The number of trucks in the traffic mix is often perceived to be a prime issue. To this end, accuracy in predicting the mix of vehicular traffic is desirable and these estimates would be of considerable use.

The number of trucks in a traffic stream may also be needed to predict noise levels and other environmental impacts.

Debates over the impacts of accommodating or allowing trucks on local streets and roads are often based more on emotion than fact. The availability of reliable estimates for truck traffic generated by specific land uses would enable quantification of potential truck traffic impacts and would be of significant benefit to all parties (e.g., general public, review agency, developer, site engineer).

A.2 Cautions

Truck trip generation data from numerous studies are presented in Section A.3 of this appendix. However, before the analyst considers using, or adapting, the data, several cautions regarding the data need to be recognized and addressed.

> **CAUTIONS**
>
> ◆ **Inconsistent definitions of trucks**
>
> ◆ **Inconsistent definitions of truck trips**
>
> ◆ **Age of the data**
>
> ◆ **Land use categories that are too broad**
>
> ◆ **Independent variables that need to be enhanced**
>
> ◆ **Potential insignificance of effects**

There are, figuratively speaking, as many definitions of the term "truck" as there are potential uses of truck trip generation data. Without a clear definition of what constitutes a truck, the compilation and analysis of truck trip generation data can be a frustrating and fruitless exercise. For example, some of the data in the literature reported in this appendix classify trucks as either courier vans, light rigid trucks, heavy rigid trucks, or articulated trucks. Another study classifies trucks as either 2- to 3-axle trucks or 4-or-more axle trucks. Another study classifies vehicles as either light-, medium-, or heavy-duty vehicles. Another study counts all vehicles with dual wheels on the rear axle as trucks. The bottom line is that local concerns regarding trucks may not match the definitions of trucks used in the studies reported in the handbook. The analyst needs to both confirm a local definition of trucks for a particular application and locate the appropriate data.

The **definition of the term "truck trips"** in some of the literature refers only to pick-up and delivery trips that use the curbside (and excludes alley deliveries). It should also be noted that throughout virtually all of the truck trip generation literature there is no differentiation (in a predictive sense) between new trips and deliveries along an established delivery route (which are akin to pass-by trips and possibly diverted linked trips, as described in Chapter 5).

Many of the **existing data sources are quite dated** (e.g., some data are more than 35 years old). The age of the data is potentially a problem for a variety of reasons:

◆ The sizes and types of trucks have changed over time;

◆ Truck usage patterns may have changed (e.g., via greater use of container trucks, use of vans and light trucks for express and high-value cargo);

◆ Delivery patterns have changed with the emergence of just-in-time delivery and intermediate distribution centers as key modes of operation in many industries;

◆ Absolute and relative volumes of freight being shipped by air, pipeline, water, rail and truck have changed;

◆ Significant state, provincial and national policy changes have affected the freight movement industry (e.g., deregulation, NAFTA); and

◆ Trip generation rates may have changed over time.

Many of the studies on truck trip generation have used **very general categories of land use**. Experience with "all-vehicle" trip generation suggests that there is considerable variation within such broad categories, so there should be a similar need for differentiation among land uses when truck studies are undertaken. The analyst is cautioned to avoid trying to apply general trip generation rates for a specific land use.

The **independent variables** that provide the greatest statistical reliability for "all-vehicle" trip generation may not be the most appropriate for estimating truck trip generation. For example, data collected at truck terminals indicate that the number of terminal doors is an appropriate predictor of truck trip generation.

While vehicle mix is of general importance with respect to some new developments and specific land uses, the problems that may occur if truck traffic is ignored (or considered as only a generic percentage of total traffic) may, in fact, sometimes be insignificant. For example, a small commercial development may only generate truck trips irregularly. The impact of such a development's truck generation on the traffic stream is most likely insignificant, often being well within the observed daily variation of trucks in the traffic stream—the addition of one or two trucks in these situations makes no realistic difference in operations.

On the other hand, larger developments, depending on absolute size and delivery patterns, may generate a significant, regular and presumably predictable volume of truck traffic. The extent and severity of the impact depends on the land use and the characteristics of the existing street/road pattern. The problem of estimating traffic impacts is to differentiate between the times when truck traffic is an important issue and when it is not.

A.3 Truck Trip Generation Experience

Section A.4 lists 23 references that present data on truck trip estimating experiences. The data contained in the documents described below hold the most promise for adaptation in estimating truck trip generation for individual development sites. However, **each cited reference has a substantial flaw in terms of its universal applicability to trucks for site trip generation purposes.**

A user of the data presented below should thoroughly review the more detailed material presented in the original documentation before simply adopting a truck trip generation rate as accurate and dependable.

Ogden (1977, 1991 and 1992)

Ogden has prepared several reports and papers that provide a comprehensive review of truck issues and contain specific information for use in estimating truck trip generation. Table A.1 is based on Australian data. The rates were presented in a 1992 paper without any information regarding sample size or any statistical measures. Although applicability in the United States may be questionable, the rates help establish a benchmark for variation in rates by vehicle and land use type.

Table A.1 Daily Truck Trip Generation Rates by Land Use (Australia) [adapted from Table 13.3 Trip Generation Rates by Land Use, Australia (1989)] Source: Ogden 1992

DEVELOPMENT TYPE	TRUCK TRIPS PER 1,000 GSF				
	FOR COURIER VANS	FOR LIGHT RIGID TRUCKS	FOR HEAVY RIGID TRUCKS	FOR ARTICULATED TRUCKS	TOTAL
Office	1.9	0.4	0	0.2	2.5
Retailing[1]					
Regional Center	0.4	0.9	0.6	0.1	2.0
Major Supermarket	0.2	0.4	0.4	0.2	1.2
Local Supermarket	0.1	0.9	0.5	0.2	1.7
Department Store	0.2	0.5	0.9	0.1	1.7
Other	0.7	0.9	0.4	0	2.0
Manufacturing	0.1	0.1	0.1	0.2	0.5
Warehouse	0.1	0	0.2	0.2	0.5
Light Industry and High Technology	1.9	0.6	0.5	0.1	3.1
Truck Depots	0.9	0.9	1.4	3.7	6.9

[1] Rate for retail is expressed in truck trips per 1,000 sq. ft. of Gross Leasable Area.

Christiansen (1979)

A definitive, but now dated, report was prepared by the Texas Transportation Institute (TTI) in 1979. This report documented many of the specific references cited in Section A.4. The report cited a data source that estimated that shopping centers in the New York City region generated approximately 1.35 daily truck stops per 10,000 sq. ft. of floor area. The data sample was small (only three centers) and covered an extremely wide range in size (between 250,000 and 5 million sq. ft.).

Reich et al. (1987)

Daily truck trips were counted at 95 suburban sites in Baltimore, MD. Land uses at the surveyed sites were classified as prepared foods, variety/pharmacy, personal services, office buildings, soft retail and retail food. Table A.2 presents the results as a function of floor area.

Cautions
- ◆ **Tremendous variation exists in observed rates**
- ◆ **Land uses do not match ITE land use codes**

Table A.2 Daily Truck Stops by Land Use (Suburban Baltimore)

LAND USE	NUMBER OF SITES	DAILY TRUCK TRIPS PER 1,000 GSF LOW — AVERAGE — HIGH
Prepared Foods	24	0.7 — 3.9 — 61.4
Variety/Pharmacy	8	0.1 — 0.6 — 10.9
Personal Services	22	0.5 — 2.3 — 5.7
Office Buildings	9	0.1 — 0.2 — 4.0
Soft Retail	14	0.4 — 2.0 — 16.7
Retail Food	18	5.2

Table A.3 Truck Trip Rates (12-Hour) per Employee in Tampa

LAND USE (NUMBER OF OBSERVATIONS)	LIGHT TRUCK TRIPS PER EMPLOYEE LOW — AVERAGE — HIGH	HEAVY TRUCK TRIPS PER EMPLOYEE LOW — AVERAGE — HIGH
Commercial (5 sites)	0.071 — 0.178 — 0.432	0.009 — 0.047 — 0.075
Office (5 sites)	0.019 — 0.038 — 0.075	0.003 — 0.009 — 0.015
Industrial (5 sites)	0.077 — 0.285 — 0.718	0.039 — 0.164 — 0.335

Gannett Fleming (1993)

Light and heavy truck trip rates were measured for retail, office and light industrial land uses in the Tampa, FL area. Rates were developed as a function of the number of employees.

Cautions

◆ Very general land use categories

◆ Rates are for a 12-hour period, not daily or peak hour

Table A.3 presents the summary, 12-hour truck trip generation rates for three general land use categories.

Tadi and Balbach (1994)

One of the more recent endeavors, which is more consistent with the development of rates and equations for specific land uses, was reported by Tadi and Balbach and concerns non-residential land uses. The data for this study were drawn exclusively from 21 locations in or near Fontana, CA, a city of about 100,000 residents located 50 miles east of Los Angeles. Regression equations are computed in the article but not shown here because they are based on only three data points (*Trip Generation* computes and presents the results of a regression analysis if the data sample is four points or greater). The

weekly trip rate tables are presented for a 24-hour period (Table A.4), morning peak hour (Table A.5), evening peak hour (Table A.6) and site peak hour (Table A.7).

Cautions

◆ Sample size for each individual use was at most three sites

◆ Conducted during period of recession in California

Table A.4 Weekday Daily Truck Trip Generation Rates (Fontana, CA)

LAND USE	INDEP. VARIABLE	2- AND 3-AXLE TRUCKS	4- TO 6-AXLE TRUCKS	ALL TRUCKS
Warehouse				
Light	1,000 GSF	0.17	0.21	0.38
Heavy	1,000 GSF	0.10	0.27	0.38
Industrial				
Light	1,000 GSF	0.33	0.27	0.60
Heavy*	1,000 GSF	0.19	0.38	0.57
Heavy*	Acre	11.90	8.63	20.53
Industrial Park	1,000 GSF	0.21	0.15	0.36
Truck Terminal	Acre	7.34	28.47	35.81
Truck Sales & Leasing	1,000 GSF	6.95	1.79	8.74

* Results based on only two data points.

Table A.5 Weekday Morning Adjacent Street Peak Hour Truck Trip Generation Rates (Fontana, CA)

LAND USE	INDEP. VARIABLE	2- AND 3-AXLE TRUCKS	4- TO 6-AXLE TRUCKS	ALL TRUCKS
Warehouse				
Light	1,000 GSF	0.01	0.02	0.03
Heavy	1,000 GSF	0.01	0.01	0.02
Industrial				
Light	1,000 GSF	0.03	0.02	0.05
Heavy*	1,000 GSF	0.00	0.02	0.02
Heavy*	Acre	0.00	0.03	0.03
Industrial Park	1,000 GSF	0.01	0.00	0.01
Truck Terminal	Acre	0.39	0.92	1.31
Truck Sales & Leasing	1,000 GSF	0.64	0.11	0.75

* Results based on only two data points.

Table A.6 Weekday Evening Adjacent Street Peak Hour Truck Trip Generation Rates (Fontana, CA)

LAND USE	INDEP. VARIABLE	2- AND 3-AXLE TRUCKS	4- TO 6-AXLE TRUCKS	ALL TRUCKS
Warehouse				
Light	1,000 GSF	0.01	0.02	0.03
Heavy	1,000 GSF	0.00	0.01	0.01
Industrial				
Light	1,000 GSF	0.01	0.00	0.01
Heavy*	1,000 GSF	0.03	0.03	0.06
Heavy*	Acre	0.58	0.08	0.66
Industrial Park	1,000 GSF	0.02	0.02	0.04
Truck Terminal	Acre	0.36	1.66	2.02
Truck Sales & Leasing	1,000 GSF	0.52	0.08	0.60

* Results based on only two data points.

Table A.7 Weekday Truck Trip Generation Rates for the Site Peak Hour[1] (Fontana, CA)

LAND USE	INDEP. VARIABLE	2- AND 3-AXLE TRUCKS	4- TO 6-AXLE TRUCKS	ALL TRUCKS
Warehouse				
Light	1,000 GSF	0.03	0.03	0.06
Heavy	1,000 GSF	0.01	0.03	0.04
Industrial				
Light	1,000 GSF	0.03	0.02	0.05
Heavy*	1,000 GSF	0.02	0.03	0.05
Heavy*	Acre	0.08	0.08	0.16
Industrial Park	1,000 GSF	0.01	0.00	0.01
Truck Terminal	Acre	0.67	1.73	2.40
Truck Sales & Leasing	1,000 GSF	1.22	0.25	1.47

* Results based on only two data points.

[1] Site peak hour is based on all trips, not just truck trips.

Truck Terminal Trip Generation, 1995 (ITE)

ITE Technical Council Committee 6A-46 undertook a review of truck trip generation at truck terminals in 1995. The committee developed trip generation materials based on contact with 19 companies. The truck terminals ranged between 800 and nearly 350,000 gross sq. ft. in size, with between 1 and 210 employees and between 0 and 113 loading dock doors.

Two regression equations (using a total sample size of 18) were developed that related weekday daily truck trips to the number of terminal doors and employees, respectively:

Weekday Daily Truck Trip Ends = $1.00 \times$ number of terminal doors + 8.96 ($R^2 = 0.69$)

Weekday Daily Truck Trip Ends = $2.06 \times$ number of terminal employees – 3.44 ($R^2 = 0.73$)

While the above equations were described as being "promising," no data plots were provided to show the fit of the lines to the data. These equations should be used with extreme caution.

Wegmann et al. (1995)

One of the more recent contributions in the area of truck trip forecasting is entitled *Characteristics of Urban Freight Systems* (CUFS). This report is a comprehensive summary of topics ranging from the aggregate characteristics of the truck fleet at the national level to the design of loading docks and similar facilities. The major point of interest is the chapter on trip rates.

Many of the trip generation results are presented without comment and, in some instances, it is impossible to judge the predictive equations or materials without returning to the original work.

In the report, Wegmann reports on daily truck trips per employee at several industrial sites (furnished by the New York Metropolitan Transportation Council). For manufacturing firms, a sample of 17 sites produced an average rate of 0.19 truck trips per employee. For air cargo operations firms (all surveyed at JFK International Airport), a sample of 18 firms produced an average rate of 0.73 truck trips per employee.

In short, this report is important because it assembles readily available information on truck trip rates. It highlights the fact that much of what is known about truck trip rates is dated information (the majority of citations, models and illustrations come from data collected in the 1970s) that has not been reevaluated for geographical or temporal validity or consistency.

French and Eck, West Virginia University (1998)

West Virginia University has conducted trip generation counts at special generators throughout the state. The counts included vehicle classification data and has enabled compilation of truck trip generation rates for each surveyed land use. Trip generation data were collected at sites covering 11 land uses, including poultry processing plants, timber processing facilities, industrial parks, superstores, mixed-retail malls, mega-medical complexes, residential subdivisions, consolidated elementary schools, junior high schools and high school, and regional jails.

Some of the data and analysis results are presented in the ITE *Compendium of Annual Meeting Technical Papers* from the 1998 Annual Meeting.

A.4 References Cited in Appendix

Christiansen, Dennis L. *Urban Transportation Planning for Goods and Services*, Final Report for the Federal Highway Administration. College Station, TX: Texas Transportation Institute, Texas A&M University, 1979.

French, J.L., and Eck, R.W. *West Virginia Special Generators Study*, Compendium of Annual Meeting Technical Papers. Washington, DC: Institute of Transportation Engineers, 1998.

Technical Memorandum No. 2, Truck/Taxi Travel Survey. Tampa, FL: Gannett Fleming Inc., July 1993.

Truck Terminal Trip Generation, Summary Report by ITE Technical Council Committee 6A-46. Washington, DC: Institute of Transportation Engineers, August 1995.

Transportation Issues Survey Summary for Sunset Park Industrial Park and JFK Air Cargo Area. New York, NY: New York Metropolitan Transportation Council, Undated.

Ogden, K.W. *Urban Goods Movement: A Guide to Policy and Planning*. Aldershot, UK: Ashgate, 1992.

Reich, Larry et al. *Baltimore Truck Trip Attraction Study*. Baltimore, MD: Department of City Planning, August 1987.

Tadi, Ramakrishna R., and Balbach, Paul. "Truck Trip Generation Characteristics of Nonresidential Land Uses." *ITE Journal*. Vol. 64, No. 7, 1994.

Wegmann, Frederick J. et al. *Characteristics of Urban Freight Systems (CUFS)*, Final Report to FHWA. Knoxville, TN: Transportation Center, University of Tennessee at Knoxville, December 1995.

Additional References

Ahrens, Gerd A. et al. *Analysis of Truck Deliveries in a Small Business District*, Transportation Research Record 637. Washington, DC: Transportation Research Board, National Academy of Sciences, 1977.

"Issues and Problems of Moving Goods in Urban Areas." *Journal of Transportation Engineering*. American Society of Civil Engineers, Vol. 115, No. 1, 1989.

A Policy on Geometric Design of Highways and Streets. Washington, DC: American Association of State Highway and Transportation Officials, 1994.

Box, Paul C. "Landfill Traffic Impact Studies." *Public Works*. September 1992, pp. 61-63.

Brogan, James D. *Development of Truck Trip-Generation Rates by Generalized Land-Use Categories*, Transportation Research Record 716. Washington, DC: Transportation Research Board, National Academy of Sciences, 1979.

Brogan, James D. *Improving Truck Trip-Generation Techniques Through Trip-End Stratification*, Transportation Research Record 771. Washington, DC: Transportation Research Board, National Academy of Sciences, 1980.

Cambridge Systematics Inc., with COMSIS Corporation and University of Wisconsin-Milwaukee. *Quick Response Freight Manual*. Washington, DC: U.S. Department of Transportation and U.S. Environmental Protection Agency, September 1996.

Chatterjee, Arun et al. "Estimating Truck Traffic for Analyzing UGM Problems and Opportunities." *ITE Journal*. Vol. 49, No. 5, 1979.

Habib, Philip A. and Nehmad, Isaac R. *Trip Generation of Curbside Pickup and Delivery*. Paper presented at the 59th Annual Meeting of the Transportation Research Board. Washington, DC, January 1980.

Habib, Philip A. *Curbside Pickup and Delivery Operations and Arterial Traffic Impacts*, Report No. FHWA/RD-80/020. Washington, DC: Federal Highway Administration, February 1981.

Ogden, K.W. "Modelling Urban Freight Generation." *Traffic Engineering and Control*. March 1977.

Ogden, K.W. "Truck Movement and Access in Urban Areas," *Journal of Transportation Engineering*." American Society of Civil Engineers, Vol. 117, No. 1, February 1991.

Spielberg, Frank, and Smith, Steven. *Service and Supply Trips at Federal Institutions in Washington, DC Area*, Transportation Research Record 834. Washington, DC: Transportation Research Board, National Academy of Sciences, 1977.

Zavattero, David A., and Weseman, Sidney. *Commercial Vehicle Trip Generation in Chicago Region*, Transportation Research Record 834. Washington, DC: Transportation Research Board, National Academy of Sciences, 1981.

Effects of Transportation Demand Management (TDM) and Transit on Trip Generation

This appendix includes material that is strictly for informational purposes. It provides no recommended practices, procedures, or guidelines.

B.1 Background

The data contained in *Trip Generation* are, by definition, from single-use developments where virtually all access is by private automobile and all parking is accommodated on site. An analyst may desire to account for the potential effects of TDM programs, of transit availability, and of small-area development patterns on site-specific trip generation rates. However, despite the reported success of a number of TDM programs, the growth of transit services in suburban areas and heightened public interest in mitigating traffic impacts, there are no statistically valid data for adjusting standard trip generation rates to reflect their effects.

B.2 Concerns Regarding Reported Experience

Available literature has extensive data on the effects of TDM programs, transit service and development patterns on travel characteristics. However, several factors confound attempts to translate data for use with site-specific trip generation estimation.

CAUTIONS REGARDING REPORTED EXPERIENCE DATA

◆ Survey-based, not count-based

◆ Commuters only (not all trips generated by a site)

◆ Not necessarily peak hour (more often total day)

◆ Very little controlled before-and-after data

◆ Trip generation reductions are employer-driven or regionally-driven, not site-driven (which is different from all other trip generation estimating applications)

Data Collection

Traditionally, the data collection needed to measure TDM/transit program effectiveness has **focused on the effects on person-trips, rather than on vehicle-trips.** This is in contrast to the numbers provided in *Trip Generation*, which are based solely on vehicle-trips. Because of this emphasis on person-trips, the preferred means of collecting the information is by survey (in order to gather information on frequency of using that travel mode, "prior-to-TDM-initiation" travel mode and other information about employment and work-hours characteristics). This survey method relies on the respondent to provide accurate information on trips made. In contrast, traditional site trip generation counts are based strictly on actual field counts of the number of vehicles entering and exiting a particular site. These counts provide the level of accuracy required in *Trip Generation;* the survey-based methods may not.

Data Focused on Commuters

The TDM/transit effectiveness surveys tend to **focus solely on commuters.** The rationale is that these programs target the travel patterns of commuters and any effects on non-commuter trips to or from a site are ancillary. However, for site trip generation purposes, the analyst must account for *all* trips to or from the site. In order to translate the commuter-only data into "all trips" the analyst must know about the trips that are non-commuter. That piece of data is rarely provided in TDM/ transit effectiveness literature but has been collected in some trip generation databases (e.g., National Cooperative Highway Research Program (NCHRP) Report 323, *Travel Characteristics at Large-Scale Suburban Activity Centers*). However, transferring those relationships between databases may reduce the validity of the overall effectiveness estimations.

Time Periods

TDM/transit effectiveness surveys are traditionally concerned with the total commuter flow to and from an employment site. Many **do not document the effects on a specific peak hour of travel** generated at the site. For example, the TDM program may only affect commuters who travel outside the peak hour.

Data Isolating TDM Effects

There are **very little controlled before-and-after data** for which the only change is the initiation of TDM programs or transit services. Traditionally, the pre-TDM mode shares are determined by survey (e.g., asking the employee how he/she commuted six months before). This method relies on the memory of the survey respondent and may not adequately account for potential bias on the part of the respondent or on the impacts of any employee turnover.

The design, initiation and operation of TDM and transit programs for which trip reductions are being sought are traditionally the responsibility of individual employers, groups of employers (e.g., through a transportation management association), or a regional or local governmental agency. Therefore, these **actions are not site-driven** which is different from all other trip generation estimating applications. There are exceptions, of course. Some site-driven measures can have a significant bearing on TDM program effectiveness (e.g., the provision of on-site services, the limitation of the on-site parking supply) while others have merely minor effects (e.g., sidewalks to neighboring sites, bus stop shelters).

The concerns over the reported experience as described above may on first inspection appear to be relatively insignificant. However, the potential error introduced by these TDM/transit factors (for the sake of argument, between 5 and 10 percent) is nearly as great as the anticipated trip reductions attributable to TDM/transit (described later in Section B.3 as 5 to 20 percent). Therefore, these data need to be used with extreme caution.

B.3 Reported Typical Experience TCRP Project B-4— Cost-Effectiveness of TDM Programs

As part of Transit Cooperative Research Program (TCRP) Project B-4, 49 employers with active TDM programs were surveyed nationwide. The primary purpose of the survey was to ascertain the costs and benefits (both perceived and actual) of TDM programs to employers. In addition, information was gathered that would enable the computation of overall reductions in the number of commuter vehicles based on the TDM programs in place. The following presents a summary of the survey results as they pertain to trip generation.

CAUTION

The magnitude of the TDM program effects is only an estimate and is not based on actual before/after counts.

Several notes of caution should be emphasized regarding the TCRP study data base.

◆ An employer survey was used to determine the number of vehicles (*not the number of vehicle-trips*) used for commuting by employees. Therefore, the "with-TDM" commute mode shares are only estimates and are **not based on actual vehicle counts.**

◆ Mode shares are **for commuters only.** The trip generation rates for non-commuter trips generated at a place of employment (e.g., visitors, deliveries, non-commute trips by on-site employees) are not included in the trip reduction estimates attributable to TDM programs.

◆ Trip reduction estimates are for commuter trips spread **throughout the day.** The values are at best suitable for an overall peak period but may not be valid estimates for a particular peak hour.

◆ To quantify the trip reduction benefits of a TDM program at an individual site, it is necessary to compare the "after" condition with the "before" condition. However, the data on **"pre-TDM" mode shares are not available.** The TCRP study assumed that the "before-TDM" baseline value should correspond to the overall mode share distribution for the surrounding area (i.e., ambient conditions), based on U.S. Bureau of the Census data.

◆ Trip reduction estimates are based on **small sample sizes** (typically 10 or fewer sites).

The following classification scheme was used in the TCRP report to categorize the many TDM programs into those that are *supportive* of persons willing to commute using an alternative travel mode, actual services that directly *enable* persons to commute using an alternative mode and financial (i.e., cash) incentives that *encourage* commuters to use an alternative travel mode.

Support Measures are measures provided by employers to foster a work environment that supports commuting by alternative modes. Support measures include employee transportation coordinators, rideshare matching, promotional activities, on-site dependent care and alternative work schedules (such as flexible work hours, compressed work weeks, staggered work hours and telecommuting).

*The surveyed TDM programs that provide only **support services** were measured to have **no effect** on the number of vehicles (not number of vehicle-trips) used by commuters.*

Transportation Services include employer-based efforts such as vanpool programs, shuttle bus service to off-site transit stations, guaranteed ride home programs and the provision of on-site showers and changing facilities.

*TDM programs that involve **transportation services** provided by the employer were measured to have a noticeable impact on the number of vehicles (not number of vehicle-trips) used by commuters (an average 8 percent reduction in the number of vehicles at the survey sites).*

Economic Incentives are any steps taken by an employer to provide a monetary incentive for employees to use an alternate travel mode. These include transit subsidies, parking fees for non-rideshare vehicles, parking discounts for rideshare vehicles and transportation allowances.

*TDM programs with **economic incentives** to not drive alone were found to reduce the number of commuter vehicles generated by an employment site (not in number of vehicle-trips) by an average of 16 percent.*

*Finally, TDM programs that **combine economic incentives with transportation services** produce the **most significant effect** on commuter vehicles (not vehicle-trips) generated by a site (an average 24 percent reduction at survey sites).*

Oregon Department of Transportation— Transportation Impact Factors

The State of Oregon sponsored a study with the intent of estimating the impacts of urban form, TDM programs and transit services on travel behavior. Tables B.1, B.2 and B.3 are extracted from that study as provided in the ITE Recommended Practice *Traditional Neighborhood Development Street Design Guidelines*, 1999.

CAUTIONS

◆ Vehicle trip reduction factors are only for commute trips (not all trips generated by a site)

◆ Vehicle trip reduction factors are for all commute trips (not just those during a peak hour)

◆ Vehicle trip reduction factors include trip reductions attributable to multi-use development

---

Table B.1 presents an estimated reduction in site vehicle trip generation for sites with no transit service and as a function of the development pattern and density, pedestrian and bicycle facilities, and other characteristics.

The analyst should note that the larger trip reduction factors are achieved with development patterns that ITE would consider multi-use (see Chapter 7 of this handbook). For example, the 7 percent reduction is associated with a "mixed-use commercial…development that includes residential units." For multi-use development sites, the guidelines and trip estimation methodology presented in Chapter 7 should be used rather than the simplified approach in Table B.1.

Table B.2 presents similar vehicle trip-reduction factors for development sites within 0.25 mile of a bus transit corridor. Vehicle trip reductions as great as 7 percent are identified for commercial uses with a relatively high floor area ratio (FAR) of two or more. Table B.3 presents similar trip reduction factors for development sites near transit centers and light rail stations.

Table B.1 Transportation Impact Factors
Development with No Transit Service

TRANSPORTATION IMPACT FACTOR	DEVELOPMENT PATTERN	DENSITY/ INTENSITY	PEDESTRIAN/ BICYCLE FACILITIES	OTHER CHARACTERISTICS	SOURCES
2% Vehicle Trip Reduction	Development located adjacent to bicycle lanes.	No specified requirements.	Bicycle lanes connect residential areas with commercial and employment uses. Provide safe and secure bicycle parking.	Distances to be traveled by bicycle should be 5 miles or less. Reasonably level terrain. Locate on streets with speed limits of 35 mph or less.	JHK, 6/95
2.5% Vehicle Trip Reduction	Mixed-use commercial and light industrial development that includes residential units. At least 30% of floor area devoted to residential uses.	For commercial and light industrial development, FAR of at least 0.75 per gross acre.	Direct, safe connections between residences and commercial/ industrial uses. Preferable if safe and secure bicycle parking is provided at commercial and light industrial uses.	Commercial uses located with minimal setbacks.	JHK, 6/95 LACMTA, 11/93
5% Vehicle Trip Reduction	Residential-oriented mixed use. At least 15% of floor area devoted to commercial uses oriented toward use by residences.	Minimum residential density of 24 dwelling units per gross acre.	Direct, safe connections between residences and commercial/ industrial uses.	Commercial uses located with minimal setbacks. Commercial includes retail and non-retail uses.	LACMTA, 11/93
7% Vehicle Trip Reduction	Mixed-use commercial and light industrial development that includes residential units. At least 30% of floor area devoted to residential uses.	For commercial and light industrial development, FAR of at least 2 per gross acre.	Direct, safe connections between residences and commercial/ industrial uses.	Commercial uses located with minimal setbacks. Commercial includes retail and non-retail uses.	LACMTA, 11/93

Source: ODOT/DLCD Transportation and Growth Management Program. Reprinted with permission.

Table B.2 Transportation Impact Factors
Development Around Bus Transit Corridors

TRANSPORTATION IMPACT FACTOR	DEVELOPMENT PATTERN	DENSITY/INTENSITY	PEDESTRIAN/BICYCLE FACILITIES	OTHER CHARACTERISTICS	SOURCES
2.5% Vehicle Trip Reduction	Locate commercial and/or light industrial uses within 0.25 mile of a bus transit corridor.	Minimum FAR of 0.75 per gross acre for commercial/light industrial development.	Direct, safe connections between commercial/industrial uses and transit stops. Preferable if safe and secure bicycle parking is provided at commercial/industrial uses.	Commercial uses located with minimal setbacks. Commercial includes retail and non-retail uses.	JHK, 6/95 LACMTA, 11/93
5% Vehicle Trip Reduction	Locate commercial and/or light industrial uses within 0.25 mile of a bus transit corridor.	Minimum FAR of 1 per gross acre for commercial/light industrial development.	Direct, safe connections between commercial/industrial uses and transit stops. Preferable if safe and secure bicycle parking is provided at commercial/industrial uses.	Commercial uses located with minimal setbacks. Commercial includes retail and non-retail uses.	JHK, 6/95 LACMTA, 11/93
5% Vehicle Trip Reduction	Locate residential development within 0.25 mile of a bus transit corridor.	Minimum residential density of 24 dwelling units per gross acre.	Direct, safe connections between residences and transit stops. Preferable if safe and secure bicycle parking is provided at the most heavily used transit stops.		LACMTA, 11/93
7% Vehicle Trip Reduction	Locate commercial and/or light industrial uses within 0.25 mile of a bus transit corridor.	Minimum FAR of 2 per gross acre for commercial/industrial development.	Direct, safe connections between commercial/industrial uses and transit stops. Preferable if safe and secure bicycle parking is provided at commercial/industrial uses.	Commercial uses located with minimal setbacks. Commercial includes retail and non-retail uses.	LACMTA, 11/93
7% Vehicle Trip Reduction	Locate residential-oriented mixed use development within 0.25 mile of a bus transit corridor. Minimum 15% of floor area devoted to commercial uses oriented toward use by residences.	Minimum residential density of 24 dwelling units per gross acre.	Direct, safe connections between commercial/industrial uses, residences and transit stops. Preferable if safe and secure bicycle parking is provided at the most heavily used transit stops.	Commercial uses should be located with minimal setbacks. Commercial includes retail and non-retail uses.	LACMTA, 11/93
10% Vehicle Trip Reduction	Locate mixed-use commercial/light industrial development that includes residential uses within 0.25 mile of a bus transit corridor. At least 30% of floor area for residential use.	Minimum FAR of 2 per gross acre for commercial/industrial development.	Direct, safe connections between commercial/industrial uses, residences and transit stops. Preferable if safe and secure bicycle parking is provided at the most heavily used transit stops.	Commercial uses located with minimal setbacks. Commercial includes retail and non-retail uses.	LACMTA, 11/93

Source: ODOT/DLCD Transportation and Growth Management Program. Reprinted with permission.

Table B.3 Transportation Impact Factors
Development Around Transit Centers and Light Rail Stations

TRANSPORTATION IMPACT FACTOR	DEVELOPMENT PATTERN	DENSITY/INTENSITY	PEDESTRIAN/BICYCLE FACILITIES	OTHER CHARACTERISTICS	SOURCES
5% Vehicle Trip Reduction	Locate commercial and/or light industrial uses within 0.25 mile of a transit center or light rail station.	Minimum FAR of 1 per gross acre for commercial/industrial development.	Direct, safe connections between commercial/industrial uses and transit center or light rail stations. Preferable if safe and secure bicycle parking is provided at commercial/industrial uses, transit centers, or light rail stations.	Commercial uses located with minimal setbacks. Commercial includes retail and non-retail uses.	JHK, 6/93 LACMTA, 11/93
10% Vehicle Trip Reduction	Locate residential development within 0.25 mile of a transit center or light rail station.	Minimum residential density of 24 dwelling units per gross acre.	Direct, safe connections between residences and transit center or light rail stations. Preferable if safe and secure bicycle parking is provided at transit centers, or light rail stations.	Commercial uses located with minimal setbacks. Commercial includes retail and non-retail uses.	LACMTA, 11/93
15% Vehicle Trip Reduction	Locate commercial and/or light industrial uses within 0.25 mile of a transit center or light rail station.	Minimum FAR of 2 per gross acre for commercial/industrial development.	Direct, safe connections between commercial/industrial uses and transit center or light rail stations. Preferable if safe and secure bicycle parking is provided at commercial/industrial uses, transit centers, or light rail stations.	Commercial uses located with minimal setbacks. Commercial includes retail and non-retail uses.	LACMTA, 11/93
15% Vehicle Trip Reduction	Locate residential-oriented mixed use development within 0.25 mile of a transit center or light rail station. Minimum 15% of floor area devoted to commercial uses oriented toward use by residences.	Minimum residential density of 24 dwelling units per gross acre.	Direct, safe connections between commercial/industrial uses, residences and transit center or light rail stations. Preferable if safe and secure bicycle parking is provided at commercial/industrial uses, transit centers, or light rail stations.	Commercial uses located with minimal setbacks. Commercial includes retail and non-retail uses.	LACMTA, 11/93
20% Vehicle Trip Reduction	Locate mixed-use commercial and light industrial development that includes non-residential uses within 0.25 mile of a transit center or light rail station. At least 30% of floor area for residential use.	Minimum FAR of 2 per gross acre for commercial/industrial development	Direct, safe connections between commercial/industrial uses, residences and transit center or light rail stations. Preferable if safe and secure bicycle parking is provided at commercial/industrial uses, transit centers, or light rail stations.	Commercial uses located with minimal setbacks. Commercial includes retail and non-retail uses.	LACMTA, 11/93

Source: ODOT/DLCD Transportation and Growth Management Program. Reprinted with permission.

B.4 References

"Cost-Effectiveness of TDM Programs." Unpublished working paper (Task 2) Transit Cooperative Research Program Project B-4. Silver Spring, MD: COMSIS Corporation, Undated.

Traditional Neighborhood Development, Street Design Guidelines, A Recommended Practice. Washington, DC: ITE Transportation Planning Council Committee 5P-8. October 1999.

Accessibility Measure and Transportation Impact Factor Study, Draft Final Report. JHK & Associates, Pacific Rim Resources, and SG Associates, for the Oregon Department of Transportation/ Oregon Department of Land Conservation and Development, Transportation and Growth Management Program. January 1996.

Hooper, K. *Travel Characteristics at Large-Scale Suburban Activity Centers*, National Cooperative Highway Research Program Report 323. Washington, DC: Transportation Research Board, National Academy of Sciences, October 1990.

Summary of Literature on Multi-Use Developments

This appendix includes material that is strictly for informational purposes. It provides no recommended practices, procedures, or guidelines.

C.1 Background

Presented below are summaries of key quantitative and qualitative findings from known databases on trip characteristics at multi-use sites. For each study, data are presented (as available) on the mix and size of land uses within the site, the level of internalization of trips within the site, overall trip generation characteristics and the level of pass-by trips. In most cases, the analyses use ITE-defined independent variables. In several cases, new variables are introduced.

1. Districtwide Trip Generation Study, Florida Department of Transportation, District IV, March 1995

The Florida Department of Transportation (FDOT) sponsored this study for two reasons: first, to develop a database that could help identify internal capture rates for multi-use development sites; and second, to develop a database from which pass-by capture rates could be established.

A summary of the characteristics of the six surveyed multi-use sites is presented in Table C.1. The sites range in area from 26 to 253 acres (with four of the sites being 72 acres or less). The office/commercial square footage ranges between 250,000 and 1.3 million sq. ft. (with three of the sites having less than 300,000 sq. ft.).

Internal Trips

The proportion of daily trips generated within the surveyed multi-use sites that were internal to the sites are listed in Table C.2. The internal capture rates ranged between 28 and 41 percent, with an average of 36 percent across the six sites.

Table C.1 Characteristics of Multi-Use Sites Surveyed by FDOT

MULTI-USE SITE	SITE SIZE (ACRES)	OFFICE (SQ. FT.)	COMMERCIAL (SQ. FT.)	HOTEL (ROOMS)	RESIDENTIAL (UNITS)
Crocker Center	26	209,000	87,000	256	0
Mizner Park	30	88,000	163,000	0	136
Galleria Area	165	137,000	1,150,000	229	722
Country Isles	61	59,000	193,000	0	368
Village Commons	72	293,000	231,000	0	317
Boca Del Mar	253	303,000	198,000	0	1,144

Table C.2 Daily Internal Capture Rates at FDOT Sites

Multi-Use Development Site	Internal Capture Rate (percentage)
Crocker Center	41
Mizner Park	40
Galleria Area	38
Country Isles	33
Village Commons	28
Boca Del Mar	33
Overall Average	**36**

Table C.3 Internal Trip Capture Rates (Percentages) by Type of Trip Maker at FDOT Sites

Trip-Maker	Crocker Center	Mizner Park	Galleria Area	Average
Users	37	38	36	37
Workers	46	49	46	47
Total	**41**	**40**	**38**	**40**

Three of the multi-use sites were further evaluated to determine the internal capture rates for different types of trip makers. As listed in Table C.3, the internal capture rates for trips made by site workers are typically higher than rates found for visitors to the site (i.e., users of the multi-use site services).

The rates by trip makers are remarkably consistent across all three sites. On average, 37 percent of user trips are internal and 47 percent of worker trips are internal to the multi-use site.

Finally, three of the multi-use sites were further evaluated to deter-mine the internal capture rates of individual land uses. Table C.4 lists the reported internal capture rates by land use/trip purpose. In general, the higher internal capture rates were reported for trips to and from banks and sit-down restaurants.

Table C.4 Internal Trip Capture Rates (Percentages) by Land Use Type at FDOT Sites

LAND USE/TRIP PURPOSE	CROCKER CENTER	MIZNER PARK	GALLERIA AREA
Office (General)	11	11	7
Office (Medical)	—	15	12
Retail	36	30	42
Restaurant (Sit-Down)	54	52	—
Restaurant (Fast)	26	—	56
Hotel	30	—	29
Bank	—	48	62
Cinema	—	23	—
Multi-Family Housing	—	11	50
Retail Mall	—	—	39

Vehicle Trip Generation

The actual vehicle trip generation rates measured at the six study sites are compared to the estimated trip generation rates based on ITE *Trip Generation*, Fifth Edition, data in Table C.5. A value of less than 1.0 indicates that the number of actual overall vehicle trips generated is less than that predicted using ITE rates.

As shown in the first column of the table, the actual number of vehicle-trips generated by a multi-use site on a daily basis is substantially less than a number predicted using ITE *Trip Generation* rates for each individual component of the site (i.e., disaggregated). In contrast, the actual trip generation on a daily basis roughly equals an estimate based on the "full-size" trip generation rates for the total square footage (or comparable independent variable) for all land uses by type within the site (i.e., aggregated). Even though a high percentage of internal trips was observed at all six sites (as documented earlier), there appears to be little effect on daily vehicle trip generation rates for the overall multi-use site.

In terms of a trip generation rate for the morning peak hour, an average of the measured rates equals the aggregated ITE *Trip Generation* rate (although the six sites demonstrated a much wider range of variability than was the case for daily trip generation). The evening peak hour trip generation rates are on average 20 percent less than the aggregate site estimate based on ITE rates. This reduction is consistent across the six study sites.

Pass-By Trips

The pass-by trip proportions, as determined through intercept surveys, are listed in Table C.6 for the six study sites. It is perhaps most telling that four of the six sites are reported to have pass-by rates between 26 and 29 percent.

Table C.5 Comparison Between Actual FDOT Vehicle Trip Generation and an Estimate from ITE *Trip Generation*

	Ratio of Actual Vehicle Trip Generation to ITE Estimate			
MULTI-USE SITE	TOTAL DAILY (DISAGGREGATED)	TOTAL DAILY (AGGREGATE)	a.m. PEAK HOUR (AGGREGATE)	p.m. PEAK HOUR (AGGREGATE)
Crocker Center	0.82	0.99	1.27	0.82
Mizner Park	1.13	1.07	0.73	0.77
Galleria Area	0.71	0.99	1.09	0.84
Country Isles	0.72	1.04	1.10	0.85
Village Commons	0.69	1.06	0.92	0.80
Boca Del Mar	0.70	0.98	1.06	0.73
Overall Average	**0.77**	**1.02**	**1.00**	**0.80**

Table C.6 Daily Pass-By Rates at FDOT Sites

MULTI-USE DEVELOPMENT SITE	DAILY PASS-BY RATE (PERCENTAGE)
Crocker Center	26
Mizner Park	29
Galleria Area	40
Country Isles	28
Village Commons	14
Boca Del Mar	29
Overall Average	**28**

2. FDOT Trip Characteristics Study of Multi-Use Developments, FDOT District IV, December 1993

This study was the predecessor of the March 1995 FDOT trip generation study. Much of the data that were collected and many relationships derived in this first study are included in the 1995 study results described above. However, the 1995 study did not report on two relationships presented in the 1993 report (summarized below).

Internal Trips

Relationships were developed for estimating internal trips as a function of the combination of two land use types in terms of residential units and office/retail square footage. Strong relationships were developed for two internal trip type categories: between residential and retail uses and between retail and retail uses. The office-retail relationship was less definitive.

The study presented a working hypothesis that the number of internal trips from one land use type (A) to another land use (B) within a multi-use site is directly proportional to the size of land use A and also proportional to the size of land use B. This suggests a functional relationship of the form:

Person Trips between A and B = Constant × Land Use A × Land Use B where:

Land Use A = total site land use of type A in residential units or per 1,000 sq. ft.,

Land Use B = total site land use of type B in residential units or per 1,000 sq. ft., and

Constant = a value that is solely a function of the two land use types.

In the equation shown above, the constant can be derived from information collected on person trips between different land use types and sizes. The derived constants are listed in Table C.7.

Table C.7 Internal Trip Coefficients for Paired Land Use Types

PAIRED LAND USES	MIDDAY PEAK PERIOD (12:00–2:00 p.m.)	EVENING PEAK PERIOD (4:00–6:00 p.m.)	DAILY
Residential/Retail	0.00082	0.00103	0.00557
Office/Retail	0.00087	0.00024	0.00232
Retail/Retail	0.01219	0.00995	0.07407

For example, application of these coefficients for a particular multi-use site with 1,144 residential dwelling units, 198,000 sq. ft. of retail and 303,000 sq. ft. of office space would yield the following results:

◆ Number of daily internal trip ends between residential and retail uses is 1,262 [0.00557 × 1,144 (residential units) × 198 (1,000 retail square footage) = 1,262]

◆ Number of daily internal trip ends between office and retail uses is 139 [0.00232 × 303 (1,000 office square footage) × 198 (1,000 retail square footage) = 139]

◆ Number of daily internal trip ends between retail and retail uses is 2,904 [0.07407 × 198 (1,000 retail square footage) × 198 (1,000 retail square footage) = 2,904]

This study also collected information on internal capture rates by time of day. Total internal capture rates for the three surveyed multi-use sites are shown in Table C.8. The estimated daily midday and evening peak period internal capture rates are quite similar. The daily internal capture rates range from 28 percent to 33 percent for the three survey sites (with an overall average of 31 percent). The midday and evening peak periods produced similar ranges for the three survey sites, 30 to 35 percent and 28 to 32 percent, respectively.

The mean values for the entire survey period shown in Table C.8 have a high degree of statistical validity. Maximum two-tailed errors calculated using the binomial distribution, with 90 percent confidence level methodology, are all less than 5 percent.

3. Trip Generation for Mixed-Use Developments, Technical Committee Report, Colorado-Wyoming Section, Institute of Transportation Engineers, January 1986.

This study determined how trip generation estimates using ITE rates compared to actual driveway counts at multi-use developments in Colorado and Wyoming. Also included were interviews to determine whether persons entering and leaving multi-use sites came there for multiple purposes. The size and mix of land uses at the eight sites with interviews are listed in Table C.9.

Table C.8 Internal Person Trip Ends by Time of Day (Percentage)

TIME PERIOD	AVERAGE RECORDED AT THREE SITES	RANGE RECORDED AT THREE SITES
Daily	31	28 – 33
Midday Peak Period (12:00–2:00 p.m.)	32	30 – 35
Evening Peak Period (4:00–6:00 p.m.)	30	28 – 32

Table C.9 Characteristics of Multi-Use Sites with Interviews

SITE	SIZE (SQ. FT.)	LAND USES
1	240,917	Retail, General Office, Government Office, Restaurants, Health Club, Bank
2	731,846	Retail, Office, Restaurants, Hotel
3	500,000	Retail, Office, Restaurants, Motel, Theaters
4	115,000	Retail, Restaurants, Hardware Store, Supermarket
5	1,000,000	Regional Mall, Retail, Restaurants, Banks, Office, Theaters
6	110,000	Retail, Theaters, Restaurants, Banks
7	95,104	Retail, Restaurants, Supermarket, Medical Office
8	300,000	Retail, Hardware, Restaurants, Supermarkets, Post Office

Table C.10 Percentages of Persons within Multi-Sites by Number of Purposes (Stops) and by Primary Destination

PRIMARY DESTINATION	NUMBER OF PURPOSES/STOPS STATED BY INTERVIEWEE		
	1 PURPOSE (%)	2 PURPOSES (%)	3+ PURPOSES (%)
Bank	83	8	9
Hardware Store	76	22	2
Supermarket	77	17	6
Theater	93	7	0
Office/Work Site	68	31	1
Small Retail Shop	73	14	13
Restaurant	85	12	3
Health Club	71	29	0
Post Office	63	24	13
Total (Average)	**77**	**16**	**7**

Internal Trips

A key piece of information collected at the interview sites was the number of trip purposes that an interviewed person accomplished on a particular trip within the site. Overall, a majority (77 percent) of the interviewees indicated that their trip involved only a single stop within the multi-use site. However, this still left a significant proportion (23 percent) who indicated they were making two or more stops within the site. Based on these interview results, the study authors estimated that 25 percent of an otherwise total number of trips were eliminated with the linking of internal trips within the eight surveyed multi-use sites.

Table C.10 presents the "number of trip purposes" data, arrayed according to the primary destination. This data gives the reader a sense for which land uses tend to generate multi-stop trips within multi-use sites. Office buildings and a post office generated the greatest number of multi-stop trips. Theaters, restaurants and banks tended to generate lower-than-average numbers of multi-stop trips within the site.

Trip Generation

Vehicle trip generation data were collected at nine sites, as described in Table C.11. During both the morning and evening peak hours for the generators within the nine multi-use sites, the actual vehicle counts were less than the calculated volumes from ITE *Trip Generation* rates. On a daily basis, six of the nine actual counts were also less.

Several of the surveyed sites are predominantly shopping centers (with some peripheral office or hotel space within the site boundaries) for which trip reduction estimates are not truly valid. Table C.12 presents the comparisons between driveway counts and ITE *Trip Generation* estimates (for each disaggregated element of the site) for the three surveyed sites that best fit the traditional view of a multi-use site. The site numbers in the table correspond to site numbers used previously in Table C.11.

The measured reduction in trips generated by the site (as an indirect and perhaps direct result of an internal capture rate) varies considerably. As shown in Table C.12, during the morning peak hour, the measured reduction at the three sites with internal trips ranged from 30 to 37 percent, with an average of 33 percent. The average reduction was 29 percent during the evening peak hour (with observed values ranging between 15 and 45 percent). Finally, on a daily basis the average reduction in vehicle trips was 13 percent (with a range between 9 and 20 percent).

Table C.11 Characteristics of Trip Generation Data Collection Sites

SITE	SIZE (SQ. FT.)	LAND USES
1	154,536	Retail, Office, Government Office, Restaurants, Health Club
2	86,381	Retail, Restaurants, Bank
3	731,846	Retail, Office, Restaurants, Hotel
4	500,000	Retail, Office, Restaurants, Motel, Theaters
5	61,198	Retail, Office
6	115,000	Retail, Restaurants, Hardware Store, Supermarket
7	1,773,500	Office, Restaurants, Bank, Hotel, Medical Office, Training Center
8	177,277	Retail, Office, Medical Office, Restaurants, Health Club, Bank, Theater, Hardware Store, Supermarket
9	95,104	Retail, Restaurants, Bank, Supermarket, Medical Office

The measured driveway volumes show vehicle trip reductions that could be considered to approximate the 25 percent drop caused by internalization of trips. It was the researchers' conclusion that most of the secondary trip purposes indicated by interviewees occur because of the availability of multiple retail outlets in close proximity to major primary destinations, such as work locations, supermarkets, banks, restaurants, hotels and theaters in multi-use developments. If the secondary destinations were not in close proximity to the primary destinations, trips to the secondary destinations would not occur or would occur at a much less frequent rate.

Table C.12 Comparison of ITE *Trip Generation* with Driveway Counts

SITE NO.	a.m. PEAK HOUR			p.m. PEAK HOUR			DAILY		
	ITE	COUNT	CHANGE	ITE	COUNT	CHANGE	ITE	COUNT	CHANGE
3	1,217	855	362 (30%)	1,491	821	670 (45%)	12,838	11,706	1,132 (9%)
4	922	640	282 (31%)	1,337	1,138	199 (15%)	15,119	13,718	1,401 (9%)
7	3,878	2,448	1,430 (37%)	4,019	2,891	1,128 (28%)	30,408	24,462	5,946 (20%)

4. Trip Generation at Special Sites, Virginia Transportation Research Council, Charlottesville, Virginia, VHTRC 84-R23, January 1984.

Driveway vehicle counts were taken at a multi-use site located in a densely developed area in the Northern Virginia suburbs of Washington, DC. The multi-use site contains 606 rental units (555 of which are located in a high-rise, the remainder being multi-level townhouse units) and approximately 64,000 sq. ft. of retail/office area (including a delicatessen, a commercial cleaning company office, two building contractor offices, restaurant, bank, hospital consulting company, direct-mail advertising firm, real estate agency, management consulting group and dentist). The site is served by transit.

Vehicle Trip Generation

Table C.13 presents a comparison between the measured trip rates at the site and the estimated trips calculated from the ITE *Trip Generation*, Fifth Edition rates.

Counts were taken (and trip generation estimates developed) for the morning and evening peak hour and the weekday daily time periods. The field-counted trips were 27 percent less than the ITE-calculated rates during the evening peak hour and 17 percent less during a 24-hour period. As has been stated in previous assessments of multi-use sites in this chapter, the reasons for this reduction could be twofold: (1) internalization of trips and (2) simple randomness of the actual trip generation rates.

Table C.13 Comparison of Actual and Counted Trip Ends

	a.m. PEAK HOUR (7:00–9:00)	p.m. PEAK HOUR (4:00–6:00)	DAILY
ITE Calculated	337	764	8,222
Field Counted	440	559	6,803
Difference from Calculated	**103 Higher** (31%)	**205 Lower** (27%)	**1,419 Lower** (17%)

Internal Trips

Trip-making at the site was only measured at its boundary. No internal counts or interviews were conducted. It is not possible to estimate internal trip rates directly from a comparison between counted and ITE-calculated vehicle trip rates. Nevertheless, all other factors being equal, it appears that the evening peak hour internal capture rate is greater than that during the morning peak hour.

5. A Trip Rate Interaction Model for Mixed Land Use Developments, University of Maryland Department of Civil Engineering (Gang-Len Chang, Chao-Hua Chen, Everett C. Carter), Maryland State Highway Administration, November 1992

The objective of this study was to develop a systematic procedure for estimating the traffic impact of multi-use developments. The recommended method from the research is based on the results of surveys at three multi-use sites. The general characteristics of the survey sites are presented in Table C.14. For the purposes of this chapter, the Cross Keys development is the most representative of a multi-use site, although it is situated in an urban setting. Burke Center more closely resembles a small town or rural village, but its trip-making characteristics are nevertheless presented below. The Reston development stretches across 20 sq. mi. and is not truly a multi-use development in the context of this handbook; its trip-making characteristics are not discussed further.

Internal Trips

The measured internal capture rates for individual land uses at two applicable survey sites are listed in Table C.15. Similar to findings in other studies, the internal capture rates are higher at office buildings for the evening peak than for the morning peak (because site workers are more likely to make secondary trips during the afternoon than in the morning). The high morning internal capture rate for the retail mall is not meaningful because it represents an inconsequential number of trips that would not typically be considered in a traffic impact analysis.

Table C.14 Characteristics of Survey Sites

	CROSS KEYS	BURKE CENTER	RESTON
Size	72 acres	1,700 acres	14,046 acres
Residences	942	19,643	56,188
Single-Purpose Office	104,841 sq. ft. (service-oriented)	17,254 sq. ft. (service-oriented)	294,000 (non-service)
Multi-Purpose Building	61,000 sq. ft. (bank, retail, office, medical)	—	847,950 sq. ft. (office, bank, retail, hotel, theater)
Retail	—	117,269 sq. ft.	—

Table C.15 Internal Trip Capture Rates at Individual Land Uses in Multi-Use Sites

	CROSS KEYS			BURKE CENTER		
	a.m. PEAK (7:00–9:00)	p.m. PEAK (4:00–5:30)	ALL DAY	a.m. PEAK (7:00–9:00)	p.m. PEAK (4:00–5:30)	ALL DAY
Single-Purpose Office (Service-Oriented)	4%	13%	8%	7%	17%	17%
Multi-Purpose Building	1%	27%	11%	—	—	—
Retail Mall	—	—	—	29%	17%	15%

The University of Maryland study reports vehicle trip generation at each survey site, but it is unclear whether or not the counts include residential areas and whether or not some vehicle movements may have been double-counted. Therefore, the results are not presented here. The University of Maryland study did not attempt to quantify pass-by trips.

6. The Brandermill PUD Traffic Generation Study, Technical Report, JHK & Associates, Alexandria, Virginia, June 1984.

Brandermill is a large, planned multi-use development (in many respects, it is a small town/village)

located approximately 10 miles southwest of Richmond, VA. At the time of the study, there were approximately 2,300 occupied dwelling units, with 180 townhouse-style condominiums and 2,120 single-family detached units. Commercial development consisted of an 82,600-sq. ft. shopping center, a 63,000-sq. ft. business park, a 14,000-sq. ft. medical center, and a 4,400-sq. ft. restaurant. There were also recreational facilities, including a golf course, tennis courts, swimming facilities and several lakeside recreation facilities. Finally, there was a day-care center, church, elementary school and middle school. The study had the overall goal of determining the on-

site (internal) and off-site (external) traffic generation at Brandermill.

Internal Trips

The split between internal and external trips was estimated on the basis of various data. As shown in Table C.16, 51 percent of the daily trips, 55 percent of the evening peak hour trips and 45 percent of the morning peak hour trips were internal to (or captured within) the multi-use site. Additionally, 46 percent of the persons employed in Brandermill also reside there.

Travel questionnaires were distributed to residences and used to measure the level of internal trip ends for home-based trips. As shown in Table C.17, roughly 35 percent of the daily home-based trips from Brandermill residences are linked with trip ends within Brandermill. More than 39 percent of the daily trip ends *to* Brandermill residences start within Brandermill. For the shopping center trips within Brandermill, roughly two-thirds of the trips originate within Brandermill during the midday and evening peak hours.

Table C.16 Split Between Internal and External Trip Ends at Brandermill

	a.m. PEAK HOUR	p.m. PEAK HOUR	DAILY
Total Generated	2,570	2,935	33,540
External Trips	1,420	1,325	16,280
Internal Trips	1,150 (45%)	1,610 (55%)	17,260 (51%)

Table C.17 Internal Trip Ends Linked with Brandermill Residences and Retail Centers

HOURS	HOME-BASED TRIPS WITH DESTINATIONS WITHIN BRANDERMILL	HOME-BASED TRIPS WITH ORIGINS WITHIN BRANDERMILL
7:00 to 9:00 a.m.	18%	51%
9:00 a.m. to 4:00 p.m.	44%	50%
4:00 to 6:00 p.m.	55%	34%
6:00 p.m. to 7:00 a.m.	41%	34%
Daily	35%	39%

HOURS	SHOPPING CENTER TRIPS WITH DESTINATIONS WITHIN BRANDERMILL	SHOPPING CENTER TRIPS WITH ORIGINS WITHIN BRANDERMILL
11:00 a.m. to 1:00 p.m.	66%	65%
4:00 to 6:00 p.m.	66%	52%

7. Travel Characteristics at Large-Scale Suburban Activity Centers, JHK & Associates, NCHRP Report 323, 1990.

The objective of the project was to develop a comprehensive database on travel characteristics for various types of large-scale, multi-use suburban activity centers (SAC). The activity centers studied were very large and had a scale very different from typical multi-use development. Therefore, the findings of this study are applicable only in major activity centers.

Data were collected at the six suburban activity centers listed in Table C.18. Following is a summary of findings pertinent to internal trips for each of the land uses listed. It is noted that "larger centers" refers to the three centers with at least 15 million sq. ft. of office/retail space, whereas "smaller centers" refers to the remaining three, which have less than 8 million square feet. A summary of some relevant relationships that were reported in NCHRP 323 is presented in Table C.19.

Table C.18 Characteristics of NCHRP Report 323 Study Sites

SUBURBAN ACTIVITY CENTER	OFFICE SPACE		RETAIL SPACE		HOTEL ROOMS	RESIDENTIAL DWELLING UNITS
	GFA	EMPLOYEES	GLA	EMPLOYEES		
Bellevue (WA)	4.7 million	12,880	3 million	6,150	1,000	N/A
South Coast Metro (Orange Co., CA)	3.5 million	10,465	4 million	6,865	1,800	2,300
Tysons Corner (Fairfax Co., VA)	17.0 million	35,020	7 million	13,355	3,100	15,000
Parkway Center (Dallas, TX)	13.0 million	39,000	2 million	3,430	1,800	206
Perimeter Center (Atlanta, GA)	13.0 million	32,500	3 million	5,150	910	2,000
Southdale (Minneapolis, MN)	4.0 million	13,700	3 million	6,155	2,200	3,000

Table C.19 Internal Trip-Making Characteristics at NCHRP 323 Study Sites

	AVERAGE	RANGE
OFFICE EMPLOYEES		
% who make an intermediate stop		
• on the way to work	10%	7 – 15%
• on the way home from work	11%	6 – 16%
% who make midday trips internal to the activity center		
• SAC with high level of professional employment[1]	—	29 – 33%
• SAC with low level of professional employment	—	20 – 23%
OFFICE VISITORS — % from within activity center		
• a.m. Peak Period		
• all SAC	—	15 – 59%
• small SAC	30%	—
• large SAC	54%	—
• p.m. Peak Period		
• all SAC	—	15 – 68%
• small SAC	33%	—
• large SAC	58%	—
REGIONAL MALLS — % trips which are internal to SAC		
• Midday		
• all SAC	37%	7 – 68%
• small SAC	23%	—
• large SAC	47%	—
• p.m. Peak Period		
• all SAC	24%	7 – 57%
• small SAC	14%	—
• large SAC	31%	—
EMPLOYED RESIDENTS — % who work within SAC		
• all	—	13 – 50%
• small SAC	27%	—
• large SAC	33%	—
HOTEL TRIPS — % internal to SAC		
• a.m. Peak Period		
• all SAC	—	13 – 53%
• small SAC	19%	—
• large SAC	37%	—
• p.m. Peak Period		
• all SAC	—	15 – 46%
• small SAC	27%	—
• large SAC	36%	—

[1] Sites with at least 60 percent of the work force in professional, technical, managerial, or administrative positions.

C.2 References

Districtwide Trip Generation Study. Walter H. Keller Inc., for the Florida Department of Transportation, District IV, March 1995.

FDOT Trip Characteristics Study of Multi-Use Developments. Tindale-Oliver and Associates, for FDOT District IV, December 1993.

*Trip Generation for Mixed Use Development*s, Technical Committee Report. Colorado-Wyoming Section: ITE, January 1986.

Trip Generation at Special Sites, VHTRC 84-R23. Charlottesville, VA: Virginia Transportation Research Council, January 1984.

Chang, G.L., Chen, C.H., and Carter, E.C. *A Trip Rate Interaction Model for Mixed Land Use Developments.* College Park, MD: University of Maryland Department of Civil Engineering, and Maryland State Highway Administration, November 1992.

The Brandermill PUD Traffic Generation Study, Technical Report. Alexandria, VA: JHK & Associates, June 1984.

Hooper, K. *Travel Characteristics at Large-Scale Suburban Activity Centers*, National Cooperative Highway Research Program Report 323. Washington, DC: Transportation Research Board, National Academy of Sciences, 1990.

APPENDIX D
Glossary

An **acre**, as defined for this report, is the total area of a development's site. The distinction between total acres and total developed acres is not always clearly defined in the site acreage reported to ITE. Therefore, caution should be used with this variable.

Adjacent street traffic includes all traffic with direct access to a development site. In some cases where the site is serviced by some form of service roadway or roadways, the adjacent street or streets would be those that lead to the service roads and thus may not actually be contiguous to the site.

The **average trip rate** is the weighted average of the number of vehicle trips or trip ends per unit of independent variable (for example, trip ends per occupied dwelling unit or employee) using a site's driveway(s).

The **weighted average rate** is calculated by dividing the sum of all trips or trip ends by the sum of all independent variable units where paired data are available. The weighted average rate is used rather than the average of the individual rates because of the variance within each data set or generating unit. Data sets with a large variance will over-influence the average rate if they are not weighted.

The **average trip rate for the peak hour of the adjacent street traffic** is the one-hour weighted average vehicle trip generation rate at a site between 7 a.m. and 9 a.m. or between 4 p.m. and 6 p.m., when the combination of its generated traffic and the traffic on the adjacent street is the highest. If the adjacent street traffic volumes are unknown, the average trip rate for the peak hour of the adjacent street represents the highest hourly vehicle trip ends generated by the site during the traditional commuting peak periods of 7 a.m. to 9 a.m. or 4 p.m. to 6 p.m. Recent studies have indicated that these time periods have expanded in some heavily populated areas.

The **a.m. and p.m. peak hour volumes of adjacent street traffic** are the highest hourly volumes of traffic on the adjacent streets during the morning and evening, respectively.

The **average trip rate for the peak hour of the generator** is the weighted average vehicle trip generation rate during the hour of highest volume of traffic entering and exiting the site during the a.m. or p.m. hours. It may or may not coincide in time or volume with the trip rate for the peak hour of the adjacent street traffic. The trip rate for the peak hour of the generator is equal to or greater than the trip rate for the peak hour between 7 a.m. and 9 a.m. or between 4 p.m. and 6 p.m.

The **average weekday vehicle trip ends** (AWDVTE) is the average 24-hour total of all vehicle trips counted to and from a study site from Monday through Friday.

The **average weekday trip rate** is the weighted weekday (Monday through Friday) average vehicle trip generation rate during a 24-hour period.

The **average Saturday trip rate** is the weighted average Saturday vehicle trip generation rate during a 24-hour period.

The **average trip rate for the Saturday peak hour of the generator** is the weighted average Saturday vehicle trip generation rate during the hour of highest volume of traffic entering and exiting a site. It may occur in the a.m. or p.m. hours.

The **average Sunday trip rate** is the weighted average Sunday vehicle trip generation rate during a 24-hour period.

The **average trip rate for the Sunday peak hour of the generator** is the weighted average Sunday vehicle trip generation rate during the hour of highest volume of traffic entering and exiting a site. It may occur in the a.m. or p.m. hours.

The **coefficient of determination (R^2)** is the percent of the variance in the number of trips associated with the variance in the size of the independent variable. If the R^2 value is 0.75, then 75 percent of the variance in the number of trips is accounted for by the variance in the size of the independent variable. As the R^2 value approaches 1.0, the better the fit; as the R^2 value approaches zero, the worse the fit. A standard formula for calculating R^2 can be found in a statistics textbook.

The **correlation coefficient (R)** is a measure of the degree of linear association between two variables. This coefficient indicates the degree to which the model estimated values account for deviations in the individual observed values of the independent variable from their mean value. Numerical magnitudes for "least squares" models range from -1 to +1, with larger absolute values representing higher degrees of linear association.

Diverted linked trips are trips attracted from the traffic volume on roadways within the vicinity of the generator but that require a diversion from that roadway to another roadway to gain access to the site. These trips could travel on highways or freeways adjacent to a generator, but without access to the generator. Diverted linked trips add traffic to streets adjacent to a site, but may not add traffic to the area's major travel routes. For an illustration of a diverted linked trip, refer to figure 5.1 in the handbook. For a description of other trip types generated by a site, refer to definitions for pass-by trips and primary trips.

An **employee** is defined as a full-time or part-time worker. The number of employees refers to the total number of persons employed at a facility, not just those in attendance at the time the study is conducted. Caution should be used with this variable, as it has not been defined in previous editions of this publication.

The **floor area ratio (FAR)** is a measure of the intensity of the use of a piece of property, determined by dividing the sum of the gross floor area of all floors of all principal buildings or structures by the total area of the parcel.

The **gross floor area (GFA)**[1] of a building is the sum (in square feet) of the area of each floor level, including cellars, basements, mezzanines, penthouses, corridors, lobbies, stores and offices, that are within the principal outside faces of exterior walls, not including architectural setbacks or projections. Included are all areas that have floor surfaces with clear standing head room (6 feet, 6 inches minimum) regardless of their use. If a ground-level area, or part thereof, within the principal outside faces of the exterior walls is not enclosed, this GFA is considered part of the overall square footage of the building. However, unroofed areas and unenclosed roofed-over spaces, except those contained within the principle outside faces of exterior walls, should be excluded from the area calculations. For purposes of trip generation calculation, the GFA of any parking garages within the building should not be included within the GFA of the entire building. The majority of the land uses in this document express trip generation in terms of GFA. In Trip Generation, the unit of measurement for office buildings is currently GFA; however, it may be desirable to also obtain data related to gross rentable area and net rentable area. With the exception of buildings containing enclosed malls or atriums, gross floor area is equal to gross leasable area and gross rentable area.

The **gross leasable area (GLA)**[2] is the total floor area designed for tenant occupancy and exclusive use, including any basements, mezzanines, or upper floors, expressed in square feet and measured from the centerline of joint partitions and from outside wall faces. For purposes of trip generation calculations, the floor area of any parking garages within the building should not be included within the GLA of the entire building. GLA is the area for which tenants pay rent; it is the

---

[1]Institute of Real Estate Management of the National Association of Realtors. *Income/Expert Analysis, Office Buildings, Downtown and Suburban*. 1985, p. 236.

[2]Urban Land Institute. *Dollars and Cents of Shopping Centers*, 1984.

area that produces income. In the retail business, GLA lends itself readily to measurement and comparison; thus, it has been adopted by the shopping center industry as its standard for statistical comparison. Accordingly, GLA is used in this report for shopping centers. For specialty retailcenters, strip centers, discount stores and freestanding retail facilities, GLA usually equals GFA.

The **gross rentable area** (GRA)[3] is computed in square feet by measuring the inside finish of permanent outer building walls or from the glass line where at least 50 percent of the outer building wall is glass. GRA includes the area within outside building walls excluding stairs, elevator shafts, flues, pipe shafts, vertical ducts, balconies and air conditioning rooms.

An independent variable is a physical, measurable and predictable unit describing the study site or generator that can be used to predict the value of the dependent variable (in this case, trip ends). Some examples of independent variables are GFA, employees, seats and dwelling units.

The **internal capture rate** is the percentage reduction applicable to the trip generation estimates for individual land uses within a multi-use site, so that the analyst can account for internal trips at the site. These reductions are applied externally to the site (i.e., at entrances, at adjacent intersections, and on adjacent roadways).

A **multi-use development** is typically a single real-estate project that consists of two or more ITE land use classifications between which trips can be made without using the off-site road system. Because of the nature of these land uses, the trip-making characteristics are interrelated, such that there are trips made among the on-site uses. This capture of trips internal to the site has the net effect of reducing vehicle trip generation between the overall development site and the external street system (compared to the total number of trips generated by comparable stand-alone sites).

The **net rentable area** (NRA)[4] is computed in square feet by measuring inside the finish of permanent outer building walls or from the glass line where at least 50 percent of the outer building wall is glass. NRA includes the area within outside building walls excluding stairs, elevator shafts, flues, pipe shafts, vertical ducts, balconies, airconditioning rooms, janitorial closets, electrical closets, washrooms, public corridors and other such rooms not actually available to tenants for their furnishings or to personnel and their enclosing walls. No deductions should be made for columns and projections necessary to the building. Typically, the NRA for office buildings is approximately equal to 85 to 90 percent of the GFA.

Pass-by trips are made as intermediate stops on the way from an origin to a primary trip destination without a route diversion. Pass-by trips are attracted from traffic passing the site on an adjacent street or roadway that offers direct access to the generator. Pass-by trips are not diverted from another roadway. Thus, impacts at the entrances and exits to proposed sites should be based on trip generation rates or equations; impacts on adjacent streets can be based on a reduced forecast to account for pass-by trips. For an illustration of a pass-by trip, refer to figure 5.1 in the handbook. For a description of other trip types generated by a site, refer to definitions for diverted linked trips and primary trips.

Primary trips are made for the specific purpose of visiting the generator. The stop at the generator is the primary reason for the trip. The trip typically goes from origin to generator and then returns to the origin. For example, a home-to-shopping-to-home combination of trips is a primary trip set. For an illustration of a primary trip, refer to figure 5.1 in the hand-

---

[3]Institute of Real Estate Management of the National Association of Realtors. *Income/Expert Analysis, Office Buildings, Downtown and Suburban.* 1985, p. 236.
[4]Ibid.

book. For a description of other trip types generated by a site, refer to definitions for pass-by trips and diverted linked trips.

A **servicing position** is defined by the number of vehicles that can be serviced simultaneously at a quick lubrication vehicle shop or other vehicle repair shop. That is, if a quick lubrication vehicle shop has one service bay that can service two vehicles at the same time, the number of serving positions would be two.

The **standard deviation** is a measure of how widely dispersed the data points are around the calculated average. The lower the standard deviation, meaning the less dispersion there is in the data, the better the data fit. In *Trip Generation*, the statistics reported are based on a "weighted average" not an "arithmetic average" and therefore, the standard deviation is an approximation.

A **trip or trip end** is a single or one-direction vehicle movement with either the origin or the destination (exiting or entering) inside a study site. For trip generation purposes, the total trip ends for a land use over a given period of time are the total of all trips entering plus all trips exiting a site during a designated time period.

A **vehicle fueling position** (VFP) is defined by the number of vehicles that can be fueled simultaneously at a service station. For example, if a service station has two fuel dispensing pumps with three hoses and grades of gasoline on each side of the pump, where only one vehicle can be fueled at a time on each side, the number of vehicle fueling positions would be four.

Index

A

Adjacent Street Traffic, 4, 81, 145
    Peak Hours, 15–16, 145
Automobile Parts Sales (Land Use 843), 51
Average Trip Rates (see Weighted Average Trip
  Generation Rate/Equation)

B

Best Fit of Data, 3, 7–8

C

Capture Rate (see Internal Trips/Internal Capture
  Rate)
Coefficient of Determination (R^2), 3, 8–9, 146
Correlation Coefficient (R), 146
Convenience Market (Open 24 Hours) (Land Use
  851), 54–55
Convenience Market with Gasoline Pumps (Land Use
  853), 56–59

D

Daily Variation, Time Period for Analysis
    Shopping Center Traffic, 6
Data Collection
    Local, 15–23
    Multi-Use Development, 101–110
    Pass-By Trips, 78–81
Data Collection Forms
    Multi-Use Development Interview Form,
      104–105
    Pass-By Trip Interview Form, 80–81
    Trip Generation Data Collection Form, 25–27
Data Plot, 7
Development Densities, 83
Discount Supermarket (Land Use 854), 60–61
Diverted Linked Trips, 29–30, 34, 36–77, 146
Drive-in Bank (Land Use 912), 65

E

Electronics Superstore (Land Use 863), 62
Employee, 146
Equation (see Regression Equation)
Estimating Trip Generation, 7–13
    Generalized Land Use, 83–84
    Multi-Use Development, 86–100
    Procedure, 9–13

F

Fast-Food Restaurant with Drive-Through Window
  (Land Use 934), 68–70
Fast-Food Restaurant with Drive-Through Window
  and No Indoor Seating (Land Use 935), 71
Floor Area Ratio (FAR), 84, 146
Free-Standing Discount Stores (Land Use 815), 37–40
Free-Standing Discount Superstore (Land Uae 813),
  36
Furniture Store (Land Use 890), 64

G

Gasoline/Service Station (Land Use 944), 72–73
Gasoline/Service Station with Convenience Market
  (Land Use 945), 74–77
Generalized Land Use (see Land Use, Generalized)
Gross Floor Area (GFA), 4, 146
Gross Leasable Area (GLA), 146
Gross Rentable Area (GRA), 147

H

Hardware/Paint Store (Land Use 816), 41
High-Turnover (Sit-Down) Restaurant (Land Use
  932), 66–67
Home Improvement Superstore (Land Use 862), 62
Hourly Variation, Time Period for Analysis
    Shopping Center Traffic, 5

I

Independent Variable, 3, 147
    Selection, 3–4, 18
Internal Trips/Internal Capture Rate
    Definition, 147
    Multi-Use Development, 86, 88, 91–103, 129–140
Interviews/Questionnaires (see Data Collection
  Forms)

L

Land Use
    Selection for Study, 16
Land Use, Pass-By Trip Tables/Figures, by ITE Land
  Use Code
    813, Free-Standing Discount Superstore, 36
    815, Free-Standing Discount Stores, 37–40
    816, Hardware/Paint Store, 41
    820, Shopping Center, 42–50

We Want Your Comments...

The Institute of Transportation Engineers (ITE) would like to know what you think about the *Trip Generation Handbook*, Second Edition, a Recommended Practice. Standards, recommended practices and informational reports published by ITE are prepared by transportation engineers and planners who have volunteered their time and effort to the project task.

ITE encourages and welcomes your views and opinions on the topics discussed in this report.

1. **Please describe any errors or inconsistencies you have noted in this report. Please note page numbers and, if possible, a copy of the page(s) containing the error. Attach additional sheets if needed.**

   Description and page(s):

   _____
   _____
   _____
   _____
   _____
   _____
   _____
   _____
   _____
   _____
   _____
   _____

2. **Are there specific subjects or issues that were not covered in this recommended practice that should have been?**

   _____
   _____
   _____
   _____
   _____
   _____
   _____
   _____

3. **Are there specific subjects or issues that were covered in this recommended practice that should not have been?**

   _____
   _____
   _____
   _____
   _____
   _____
   _____
   _____

4. Do you have any other suggestions or comments regarding this recommended practice?

_____
_____
_____
_____
_____
_____
_____

5. Do you know of any additional data or information that can be made available to ITE to help improve the content of this report?

_____
_____
_____
_____
_____
_____
_____

The following information is optional:

Name _____

Title _____

Agency or Firm _____

Address _____

City _____

State/Province_____ Postal Code_____

Country _____

Telephone _____ Fax _____ E-mail _____

Thank you!

Please return this form to:
Institute of Transportation Engineers
Technical Projects Division
1099 14th Street, NW, Suite 300 West
Washington, DC 20005-3438 USA
Telephone: +1 202-289-0222
Fax: +1 202-289-7722
ITE on the Web: www.ite.org